Basics of Pharmaceutical Manufacturing and Quality Operations

This book provides guidance on how to meet the requirements of the pharmaceutical industry as a beginner. It includes procedures for production and packaging, batch auditing as well as all quality measures used in the pharmaceutical industry. This book also provides questions and answers with each chapter for institutes and trainers providing basic training to the new graduates and newcomers to the industry.

Basics of Pharmaceutical Manufacturing and Quality Operations: A Comprehensive Guide is primarily written for anyone in the pharmaceutical industry interested in development and manufacturing of active pharmaceutical ingredient (API) and finished pharmaceutical manufacturers in both sterile and non-sterile areas. The book is a simple, concise, and easy to use reference tool covering basic quality concepts required by the pharmaceutical educational institutions and professional certification bodies. It describes details of all GXP activities that are directly related to quality, safety, and efficacy of the products manufactured under the umbrella of Quality Operations, common testing methods which are used in any modern industry, Requirements of Validation and Qualification of equipment, facilities and processes, integral segments of Drug product manufacturing, storage, and distribution practices. The material provides stepwise guidance on how to evaluate, audit, qualify, and approve a pharmaceutical product to enhance the GMP within the industry.

The book is written with the idea of providing basic knowledge to undergraduate students who are preparing to enter the industry at the end of their graduation. The book would also be beneficial for institutions conducting pharmaceutical technology study courses in terms of GMP and GLP applications.

Features:

1. Provides readers and frontline healthcare product manufacturers, all the information they need to know to develop a GMP oriented industry with trained and skilled personnel and manufacture products that meet GMP and regulatory requirements.

2. Provides stepwise guidance on how to evaluate, audit, qualify, and approve a pharmaceutical product and packaging material to enhance the GMP within the industry.
3. Includes significant processes and steps in production for all common dosage forms.
4. Explains how in-process and finished products are released.
5. Provides an ideal and effective tool for anyone starting quality assurance/quality control/production responsibilities.

Basics of Pharmaceutical Manufacturing and Quality Operations

A Comprehensive Guide

Erfan Syed Asif, Ph.D
and Shahid Bader Usmani, Ph.D

CRC Press
Taylor & Francis Group
Boca Raton London New York

CRC Press is an imprint of the
Taylor & Francis Group, an **informa** business

Designed cover image: © Shutterstock and istockphoto

First edition published 2024
by CRC Press
2385 NW Executive Center Drive, Suite 320, Boca Raton FL 33431

and by CRC Press
4 Park Square, Milton Park, Abingdon, Oxon, OX14 4RN

CRC Press is an imprint of Taylor & Francis Group, LLC

ISBN: 9781032432052 (hbk)
ISBN: 9781032432465 (pbk)
ISBN: 9781003366393 (ebk)

DOI: 10.1201/9781003366393

Typeset in Times
by codeMantra

To the authors' Alma-Mater, ***Hussain Ebrahim Jamal Research Institute of Chemistry, University of Karachi,*** *which served the pharmaceutical industry not just in Pakistan but all over the world by producing scientists, inventors and innovative thinkers in Research & Development, manufacturing, and quality operations.*

Contents

About the Authors ... ix
Preface ... xi

Chapter 1 Good Manufacturing Practice (GMP), Good Lab Practice (GLP) and Good Documentation Practice (GDP) ... 1

Chapter 2 Concept of Calibration ... 15

Chapter 3 Reference Standards and Reagents ... 24

Chapter 4 Chemical Tests ... 32

Chapter 5 Physical Testing of Raw Materials, Intermediate, and Finished Products ... 49

Chapter 6 Data Integrity ... 63

Chapter 7 Process Validation ... 68

Chapter 7.1 Protocol Generation ... 73

Chapter 7.2 Sampling and Testing Plan ... 87

Chapter 7.3 Process Validation Report Writing ... 93

Chapter 8 Cleaning Validation ... 117

Chapter 9 Method Development and Method Validation ... 126

Chapter 10 Qualification and Its Segments ... 134

Chapter 10.1 User Requirement Specification 160

Chapter 10.2 Functional Specification 175

Chapter 11 Stability Studies, Stability Indicating, and Methods Degradates 182

Chapter 12 Technology Transfer 189

Chapter 13 Dispensing....193

Chapter 14 Production....198

Chapter 15 Packaging....209

Chapter 16 Role of Quality Assurance....214

Chapter 17 Role of Quality Control 218

Chapter 18 Logistics – Cold Chain Supply 226

Chapter 19 GMP for Biologics and Similar Medicines....230

Chapter 20 QA/QC Release....236

Index....251

About the Authors

Erfan Syed Asif earned his PhD in organic chemistry and has expertise in various areas of quality operations with over 28 years of experience in pharmaceutical industries in Pakistan, UAE, Kingdom of Saudi Arabia, United States, and Canada. He has worked in U.S. FDA- and Health Canada-approved facilities in various Leading roles. He has extensive experience in overseeing qualification projects for manufacturing equipment, utilities, systems, sterilization techniques, aseptic processes simulation, API site audits and sterile and non-sterile products manufacturing processes, and decommissioning of equipment.

Dr. Asif currently holds the position of senior quality consultant with PharmEng Technology Inc, Toronto, Canada and is overseeing various Quality related projects at Sanofi-Pasteur, Toronto, Canada.

Dr. Asif is the author of many research publications in internationally published chemistry journals and conference proceedings. He has authored and co-authored four books all published by Taylor & Francis/CRC Press.

Dr. Asif also participates as a guest speaker at Windsor University, Ontario, Canada to deliver lectures on GMP to the graduate students of Masters in Medical Technology, International Center of Chemical and Biological Sciences (ICCBS), Karachi Pakistan and Association of Pakistani Physicians in North America (APPNA).

Dr. Asif is a certified ASQ Quality Auditor and a member of American Society for Quality as well as a member of Royal Society of Chemistry, UK.

Usmani Shahid Bader obtained his Ph.D. in organic chemistry and has expertise in various areas of quality operations with over 25 years of experience in pharmaceutical industries in Pakistan and Canada.

He has worked in a leading role in U.S. FDA- and Health Canada-approved facilities. He has extensive experience in overseeing quality projects.

He has 33 publications in leading international journals. Due to his extensive research contribution, he was selected to conduct one year of research at the Tokyo Institute of Chemistry, Japan, among 1 of 14 scientists selected from all over the world.

Dr. Usmani currently holds the position of Group Leader with Viatris Pharmaceutical Canada, Toronto.

Preface

Pharmaceutical, medical, and biotech industries are regulated worldwide to be in compliance with current good manufacturing practice (cGMP). The GMP starts right from the procurement of raw and packaging material and continues through to the testing of raw and finished products up to the release of products for commercial use. Every pharmaceutical company is required to create and adhere to their standards and regulatory requirements. To understand these standards before stepping into this regulated industry is always a preferred way.

Looking into the nature of this highly regulated industry, the handbook of Introduction to Pharmaceutical Industry has been organized as a source to provide the initial basic education to the beginners in the industry as well as the students who are seeking the diplomas in pharmaceutical technology.

Over the past few decades, there has been enormous progress, changes, addition and enhancement over the concept and approach of Good Manufacturing Practices including the most recent idea of introducing Quality by design (also called QbD). However, all the basic elements of GxP have not been explained in detail together in one book.

This book provides complete information about the GxP measures taken in different dosage forms and manufacturing operations starting from dispensing up to QC/QA release of finished products detailing all the necessary actions following the GxP. This volume will serve as a valuable reference guide that will be referred again and again.

No other book in print deals exclusively with training for the Good Manufacturing Practices in Pharmaceutical and biotechnology industries.

This book is a good reference for entry-level Laboratory and Production technicians, managers, supervisors, and scientists in the pharmaceutical industry. The primary intent of this work is to guide Pharmaceutical personnel in the production of medicines

The contents of the book are intended to build quality into the routine operations to comply with regulatory requirements. We strongly believe that the training department of a pharmaceutical company also can use the contents to train the new comers in the Quality control and Quality Assurance, Research and development, Compliance, and Material management staff so as to provide them a base to understand the quality and GxP concepts before they get into routine operations. Besides, this book is especially a valuable tool for the colleges, institutes, and even universities providing the pharmacy and pharmaceutical technology education, degrees, and/or diplomas. The ready-to-use questions and answers provide a good source for experienced trainers and teachers to modify them and use to check the knowledge of the trainees and students.

This book takes into account all major international regulations such as the Food and Drug Administration (FDA), European Medicines Agency (EMEA), the International Conference on Harmonization (ICH) guidelines, and the industry standard ISO

9000, so as to be in compliance with the GMP guidelines internationally. No other book in print deals exclusively with all GMP factors together. The introduction to pharmaceutical Industry can be used directly or tailored to achieve a comprehensive process for the training of new employees and pharmaceutical technology students.

It provides readers and front-line biotech, APIs, products manufacturers, R&D management, and Material management with all the information needed to have a comprehensive training program on GMP.

- It is a simple, concise, and easy-to-use reference tool covering basic concepts and all elements of Good Manufacturing Practices.
- The text is a valuable timesaver for companies that are in the process of developing a training program and department to bring consistency in their operations.
- The elements provided in this book can be easily tailored to incorporate changes to add in-house requirements for training.
- The topics provide stepwise guidance on how to explain new and existing staff on the concepts of GxP and increase awareness.

We strongly believe that a facility's Quality control and Quality Assurance, Research and development, Compliance and Material management staff should avail themselves of these elements to ensure that particular needs in the course of GxP training are addressed with reference to the regulatory requirements. It is a good reference for entry-level QA technicians, managers, supervisors, and scientists in the pharmaceutical industry. The primary intent of this work is to guide the beginners in any department of operations in the Pharmaceutical industry and Quality Compliance professionals in the identification and provision of trainings.

Some of the reasons for purchasing this book are:

While explaining the details of Equipment Qualification and process Validation, questions have been created as a workshop for the students to generate an IQ or OQ protocol for a given equipment. In the answer complete templates have been provided to explain how to generate an Installation or operational

Qualification protocol, a process validation protocol, a validation report, User Requirements Specification as well as Functional Requirements Specification in the form of templates so that it is easy for the beginners to understand by looking at the examples instead of only reading theoretical contents.

The materials have been derived from the basic principles of GMP, from supported reference materials, and also from the personal experiences of the authors, who have been involved in GMP and Quality operations over the last 27 years in the pharmaceutical industries. Considerable care, thought, guides, and learning elements have been put together in this manual.

1 Good Manufacturing Practice (GMP), Good Lab Practice (GLP) and Good Documentation Practice (GDP)

Good manufacturing practice (GMP) is the norms and practices that regulate pharmaceuticals and food products manufacturing processes in every single step of pharmaceutical operations. These practices and norms are collectively known as Good manufacturing practices or GMP. Following GMP ensures that the products are safe, effective, and high-quality, protecting public health. The regulated practices in the laboratory are known as good laboratory practices or GLP. While practicing the regulations while preparing a document in the pharmaceutical industry is known as good documentation practices or GDP. Collectively, all these practices are abbreviated as GxP, where 'x' represents manufacturing, laboratory and documentation, respectively.

This chapter discusses the key principles of GxP and the importance of compliance with these guidelines.

First of all, we discuss GMP. The key principles of GMP include:

QUALITY CONTROL

The manufacturing process should be designed to ensure that the final product meets the desired quality standards. Quality control measures should be put in place and implemented in the production process to identify and correct any defects.

DOCUMENTATION

Detailed documentation should be maintained throughout the manufacturing process to ensure that every step is recorded and can be traced back if necessary. This includes standard operating procedures, batch records, and test results.

EQUIPMENT

Manufacturing equipment should be regularly calibrated and maintained to ensure that it functions correctly and produces the desired results.

DOI: 10.1201/9781003366393-1

PERSONNEL

All personnel involved in the manufacturing process should be appropriately trained and qualified to perform their tasks. They should also follow standard operating procedures to ensure consistency and quality.

FACILITIES

The manufacturing facilities should be designed and maintained to provide a suitable product production environment. This includes controlling factors such as temperature, humidity, and air quality.

MATERIALS

Raw materials and finished products should be stored, handled, and transported in a way that prevents contamination or damage.

VALIDATION

The manufacturing process should be validated to ensure it can consistently produce a product that meets the desired quality standards.

Compliance with GMP guidelines is essential to ensure that products are safe, effective, and high quality. Failure to comply with these guidelines can result in product recalls, regulatory sanctions, and damage to a company's reputation. In addition, compliance with GMP guidelines can increase efficiency and reduce costs by minimizing waste, improving productivity, and reducing the need for rework or corrective actions.

In conclusion, GMP is a set of guidelines ensuring that pharmaceuticals, medical devices, and food products are safe, effective, and high quality. Compliance with GMP guidelines is essential to protect public health and ensure the success of a company's operations. By following the key principles of GMP, manufacturers can improve efficiency, reduce costs, and maintain their reputation as a provider of high-quality products.

Any industry should have a well-designed and correctly implemented pharmaceutical quality system that incorporates GMP.

The quality department should ensure that the entire pharmaceutical quality system is resourced with qualified personnel and suitable/sufficient premises, equipment, and facilities.

The basic concepts of quality management, GMP, and quality risk management are interrelated.

PHARMACEUTICAL QUALITY SYSTEM

The quality system covers all matters that influence the quality of a drug. It is the total of the arrangements to ensure that drugs are of the quality required for their intended use. It incorporates GMP.

GMP applies to all drug product lifecycle stages: from the manufacture of drugs to technology transfer, commercial manufacturing, distribution, and discontinuation of the product.

The pharmaceutical quality system can even extend innovation and continual improvement of the products.

To ensure whether the pharmaceutical quality system is properly set up for fabricating, packaging, labeling, testing, distributing, importing, or wholesaling drugs, the following steps must be taken:

Design, plan, implement, maintain, and continuously improve the system to allow the consistent delivery of products with proper quality attributes.

- Manage product and process knowledge throughout all lifecycle stages.
- Design and develop drugs in a way that takes into account GMP requirements.
- Clearly outline management responsibilities.

Make arrangements for:

- Manufacturing, supplying, and using the correct starting and packaging materials.
- Selecting and monitoring suppliers.
- Verifying that each delivery is from the approved supply chain.
- Ensure processes are in place to properly manage outsourced activities.
- Establish and maintain control by developing and using effective monitoring and control systems for process performance and product quality.
- Records are appropriately reviewed and protected from accidental or intentional modification or deletion.

Quality should implement a data governance system.

Consider the results of product and process monitoring in batch release and in the investigation of deviations to preventive action to avoid potential deviations in the future.

Carry out all needed controls on intermediate products and any other in-process controls and validations.

Ensure continual improvement by making quality improvements appropriate to the current process and product knowledge level.

Make arrangements to evaluate and approve planned changes before implementing them. Regulatory notification and approval of changes must be done as required.

After implementing any change, conduct an evaluation to confirm that your quality objectives were achieved. Ensure there was no unintended negative impact on product quality at the time of release and through its shelf life.

Apply an appropriate level of root cause analysis when investigating deviations, suspected product defects and other problems. This can be determined using quality risk management principles. In cases where the true root cause(s) not be determined, identify the most likely root cause(s) and address those.

Where human error is suspected or identified as the cause, this should be justified with objective evidence. Ensure that process, procedural or system-based errors or problems have not been overlooked, if present.

Determine the full impact of the deviation, and document how you reached your conclusion.

Identify and carry out appropriate corrective actions and/or preventive actions in response to investigations. Monitor and assess the effectiveness of such actions in line with quality risk management principles.

Ensure the Quality Control Department approves each production batch of drugs before you sell or supply them.

Quality must produce and control drugs according to marketing authorization requirements and any other regulations relevant to drug production, control and release.

Ensure that drugs and the materials that go into making and packaging the stored, distributed and handled properly, and quality is maintained throughout their shelf life.

Implement a process for self-inspection and/or quality audit regularly and assess the effectiveness and applicability of the pharmaceutical quality system.

Senior management should participate actively in the pharmaceutical quality system. They should ensure an effective pharmaceutical quality system is in place. This includes making sure the system is properly resourced and that roles, responsibilities and authorities are defined, communicated and implemented throughout your organization. Senior management should also ensure staff—at all levels and sites within your organization—support and are committed to the pharmaceutical quality system.

Have senior management periodically conduct a management review of pharmaceutical quality system operations to continually identify risks and opportunities to improve products, processes and the system itself.

Define and document your pharmaceutical quality system. A quality manual or equivalent documentation should describe the system, including management responsibilities.

Good Manufacturing Practices for Drugs

GMP are part of quality assurance. Quality assurance that drugs are consistently produced and controlled. Drugs must meet the quality standards for their intended use as per product specifications.

GMP is concerned with both production and quality control. To meet GMP requirements, you must:

- Clearly define all manufacturing processes. Review them systematically. Provide evidence that they can consistently manufacture the required quality that complies with their specifications.
- Critical steps of manufacturing processes and key changes to the process must be validated.
- The staff must be qualified and trained. Proper hygiene and sanitization must be followed.
- Adequate premises and space must be used.

Suitable equipment and services must be provided.
Correct materials must be used; containers must be properly labeled.
The material must be well segregated, making sure that there is no mix-up in materials.
Approved procedures and instructions must be provided.
The material must be stored and transported in a suitable environment.
Step-by-step instructions and procedures must be provided in clear and direct language.
Operators must be properly trained to carry out procedures. Ensure they understand the importance of meeting GMP requirements as part of their role in assuring patient safety.
Records of all manufacturing processes must be created. Record the quantity and quality of all components used during the process.
If any deviation is made, document that, and investigate significant deviations to determine the root cause and impact. Ensure proper corrective and preventive action is taken.
Records of fabrication, packaging, labeling, testing, distribution, importation and wholesaling must be kept. In this way, a complete history of a lot can be traced.
Handling and transportation of drugs and their ingredients must be done in a controlled environment to minimize any risk to their quality.
There must be a system in place for recalling drugs from the sales market.
Market complaints must be examined about drugs. Investigation must be done of quality defects, and appropriate measures must be taken to prevent problems.
At a time, only one product must be produced to avoid any mix-up of materials.

Good Laboratory Practices (GLP/GMP for Quality Control)

The role of quality control is:

- sampling
- specifications
- testing
- documentation
- release procedures

All the raw materials, packaging materials and products for use or sale must be released if their quality is satisfactory. Quality control ensures that the company carries the necessary and relevant tests.

Here are the basic quality control requirements:

A facility with trained personnel and approved procedures for sampling and testing of raw materials, packaging materials, and intermediate bulk and finished products. The facility must be able to monitor environmental conditions.

An approved system in place to sample raw materials, packaging materials and intermediate, bulk and finished products using authorized personnel and approved methods. Methods must be validated. The equipment, and instruments must be calibrated, and computer systems must be validated.

All manual and/or electronic records must be kept showing all required sampling, inspecting and testing procedures. If any deviations are made, investigate, and document the incident.

Ensure finished products contain active ingredients complying with the qualitative and quantitative composition stated in the approved documents. Ensure they are of the purity required, enclosed within their proper containers, and correctly labeled and stored.

Results of inspection and testing of intermediate, bulk and finished products and materials against specification must be documented.

Ensure each raw material used to produce a drug is covered by specifications. These specifications must be approved and dated by the person in charge of the quality control department or by a designated alternate.

Review product release procedures and evaluation of relevant production documentation, as well as an assessment of deviations from specified procedures.

The product cannot be released for sale before they are approved by your quality control department.

A sufficient portion of the raw material and finished product sample must be stored in a controlled condition to allow future examination if needed.

Good Documentation Practice

Good documentation practice is the foundation of all quality systems because clear, complete, and accurate records are essential to all operations and procedures.

PRINCIPLES OF GOOD DOCUMENTATION

All steps related to the manufacturing, testing, packing, products, APIs, excipients, food ingredients, and medical devices should be documented.

Good documentation principles for manual or electronic records include the following, as applicable:

Records should be clear, concise, accurate, and legible. Data entries should be recorded promptly when actions are performed. If something is not documented, it means it is not done.

Backdating and postdating are not allowed. All corrections to the original entries should be initialled and dated. All the corrections must be properly explained.

Data entries should be traceable to the worker who made the entry. Uncommon abbreviations and acronyms should be defined. Controls should be in place to protect the integrity of the records.

If ink may have faded over time (e.g., thermal paper), a copy can be used to verify its accuracy; the copy should be initialled and dated.

Notebooks, data sheets, and worksheets should be traceable. An adequate documentation system is needed to ensure data integrity and the availability of current and archived records.

Records should be retained per regulatory requirements and be readable during the retention period.

All pages should be numbered. Attachments or any other supporting documents should be numbered with reference to the parent document.

DATA RECORDING

Paper forms, data sheets, and worksheets can be used for documentation. Notebooks and logbooks should be maintained. Instrument printouts must be kept with data.

Electronic data obtained with a system such as an electronic data system, laboratory information management system (LIMS), or electronic laboratory notebook must be kept with documents.

All data should be permanently recorded directly and legibly during the activity. It should be recorded in indelible ink if it is a paper record. All data entries should be traceable to the worker who made the entry and when it was recorded.

Any change to an entry should be made in a way that does not obscure the original entry, with an explanation in cases where the reason for the change is not obvious. Changes should be traceable to who and when the change was made. For clarity, predefined correction codes may be used; for example, WD = wrong date, EE – Entry error, and CE = Calculation error.

Notebook pages and worksheets should be used consecutively, and information should be recorded chronologically. GMP records such as batch records, test methods, and specifications should be given unique identifiers and use version control for the documents.

All data entry fields should be completed. A single line and/or "N/A" should be drawn through a part of a page that is not used. If the record is in an electronic system and the system provides traceability of who filled each field and when the field can be left blank.

All dates should be expressed in a format that indicates the day, month, and year.

All GMP records for data collection should undergo appropriate review and signature by a second person to confirm the entries, accuracy, and compliance. Additional signatures may be required on the basis of local standard operating procedures (SOPs) for different levels of review (e.g., performed by, verified by, checked by, reviewed by, and approved by) as accountability steps.

An official record of signature and initials should exist for each employee or can be included within the document. Controls should be in place for assigning signature approval requirements and delegation of signature authority when needed.

In the event that verified copies of raw data have been prepared, the verified copy may be substituted for the original source as raw data.

All multiple-page data sheets or instrument printouts in paper form should be signed/initialled on the first or last page with a note indicating the total number of pages. The first and subsequent pages should be uniquely identified to the activity being performed, such as the notebook reference, study number, or worksheet reference.

TYPES OF GMP DOCUMENTS

The following documents should include the following information.

Laboratory Records

Laboratory records should be organized to ensure that the records are concise, clear, legible, and accurate and detail the following:

Description of materials, such as reagents. This information includes the material name, manufacturer and lot number, concentration, expiration date, grade and a reference to the lab notebook if prepared in the lab.

Identification of equipment used. This information typically includes the equipment name, unique control number, and calibration expiration date, as applicable.

Documents should include; Procedures used, formulae and calculations and results and conclusions.

Documentation Related to Instruments

All equipment used in manufacturing, testing, packing, or raw material, API, finished product, or another similar item should be maintained and qualified for its intended use. The documentation related to equipment includes:

Procedures for operation, equipment use, maintenance records and calibration records

Deviations and Investigations

All deviations and errors must be documented. The deviation should be evaluated and investigated as appropriate. Procedures should be in place for documenting, evaluating, and investigating such events.

Documentation of the investigation should include a description of the event, root cause investigation, evaluation of data trend, responsibilities of people involved in the investigation or deviations.

In addition, the impact should be assessed, and corrective action and preventive action (CAPA) should be in place.

Batch Records

A template for the manufacture of a specific product should be created. This template should be used to document the steps and materials involved in producing a specific batch. The management should approve this document for the manufacturing site or packaging site:

This document should include the product name, batch number, and manufacturing site.

This document should include product name, batch number, manufacturing site, manufacturing process, Target weights of raw materials, condition of production, In-process sampling or testing

Certificate of Analysis

Certificate of Analysis includes a report of analytical testing results and specifications of a particular batch. It may be a raw material, finished product or similar item.

It should also include the name of the testing side, product name and strength, the procedure used, data source, expiry and retest information and approval signature.

Standard Operating Procedures

The purpose of an SOP is to provide guidelines to the personnel who will perform a certain task. SOPs should be clear and concise.

The SOP should contain the scope, procedure, responsibilities and roles of workers, equipment to be used, approval and revision history.

Protocols and Reports

Many tasks are performed based on the preapproved protocol.

For examples:

- Equipment qualification
- Analytical method validation or verification
- Manufacturing process validation
- Analytical method or manufacturing technology transfer
- Cleaning validation
- Stability study or testing
- Comparability study

Analytical Procedures

Analytical procedures provide direction to an operator on how to perform a given analytical test.

An analytical procedure contains the purpose, test information, product information, materials and equipment, procedure, as applicable, System suitability requirements, solutions and reagents preparation, standards and samples preparation, instrument parameters, calculations, review and approval with approval dates and revision history.

TRAINING DOCUMENTATION

Personnel should be trained to perform their assigned tasks. The training should be documented, and the training records should be retained and kept readily accessible.

The training documentation should include a training description including the name of training, version, mode (self-training or instructor-led), completion date and information on the trainer.

RETENTION OF DOCUMENTS

An adequate record retention and archiving policy should be established for the above records. The required length of time depends on the regulatory requirements or company procedures; however, it should be at least one year after the batch expiration date.

CHANGE CONTROL

Change is inevitable, but that doesn't mean it is easy to manage if the outdated or manual processes are still in use. Changes can be temporary or permanent, but they can't be avoided if a compliant quality or manufacturing environment is intended to maintain.

How do we see the changes and how are they controlled in pharmaceutical industry? Change control in the pharmaceutical industry is used to modify any process or pharmaceutical procedure. It ensures that the change does not affect the output's intent and fulfills all the quality principles associated with that specific product or process.

A company may need to make changes to a process or procedure for reasons including changes to fundamental legal requirements, scientific/technical development, business constraints, or other reasons. They may have to redefine, enhance, amend or cancel requirements over and over again. Whenever a change is made to a process or procedure, it is reported by change control procedure and that is approved by the company authority.

The glossary to Annex 15 of the EU GMP Guidelines defines "change control" as: "A formal system by which qualified representatives of appropriate disciplines review proposed or actual changes that might affect the validated status of facilities, systems, equipment or processes."

By keeping in view the aforementioned statement, one must understand that to change an established process or system is not a routine process. It requires lots of commitments, understanding and considerations to ensure that the already established and validated process, if changed, is not going to jeopardize the products quality efficacy and safety. In short, it is a regulatory requirement that all changes are controlled.

Here we must understand a basic point that change control is not required only for something already existing. Even for an entirely new system, process or product, a change control is necessary. The logic behind a change control for a new entity is that once it comes into being, it causes a change in the system or environment that was existing before its creation.

Why Is Change Control Important?

Change control is important for the following reasons:

1. To assure that changes do not impact Product Identity, Strength, Quality, Purity, Potency
2. Does not affect validated Processes/Systems

3. No impact on Regulatory Commitments (e.g., NDA/ANDA/BLA)
4. To provide a single, easy to access the source of information regarding changes for

    - Audits/Inspections
    - Annual product reviews (internal)
    - Annual reports (to FDA)

    Apart from the above, there are more reasons that regulatory bodies enforce a change control system to the pharmaceutical industries, which are;

    1. Change control helps ensure that the product is free of side effects when it is used by the consumer (safety)
    2. The product is exactly what the label and associated materials say it is (Identity)
    3. The product delivers the right dosage and strength over its shelf life (Strength)
    4. The product is free from physical, biological, and chemical contamination (Purity).
    5. Manufacturers establish and follow quality systems to help ensure that their products consistently meet applicable requirements and specifications (Quality).

    By adhering to the Change Control policy and procedure, companies achieve the following:

5. Manufacture a product that meets the specification
6. Keep equipment free from failure
7. Ensure sterility assurance
8. Comply with cGMP requirements
9. Prevent product quality problems including FDA 483 findings, warning letters, sub-standard products, or product recall.

Once a new system or process is to be introduced and the final decision has been taken, a formal change control request is initiated and the entire process is routed as follows;

The change control request is always initiated by the owner department or the department which is going to be the end user of the system or the process.

- Owner department will develop an implementation plan for the change/s.
- Complete the Change Control Request Form.
- State description of proposed change/s on the CCR. Provide necessary supporting documentation and data.
- Obtain approval for the change/s from the applicable Department Manager.
- Manager will determine the category of the change if it is Major or Minor.
- If acceptable, sign and date the approval section of the CCR.

- Verify that the proposed change and its implementation plan are acceptable.
- Forward the CCR to the Quality Assurance Department upon approval from Department Manager.

Quality Assurance Specialist

- Quality Assurance Specialist or Manager will assign a Change Control Tracking Number to the CCR.
- Forward a copy of the CCR to relevant personnel for review and comments.
- Subject expert/s communicates any concerns to the Quality Assurance Department.
- QA specialist will collect all comments from the subject expert and will ensure that all comments have been addressed.
- Upon addressing all concerns, the CCR shall be approved by the subject expert/s.
- Quality Assurance will subsequently approve the CCR by completing the CCR Closure, thus authorizing the applicable department to proceed with the change/s. Document and file the completed CCR.
- Initiator will than proceed with change/s upon receiving approval from Quality Assurance.
- Once all the changes have been implemented, the change control will be reviewed by QA for verification and will then be closed.

QUESTIONS & ANSWERS

1. **What is GMP?**
   Answer: Good Manufacturing Practice (GMP) regulates pharmaceuticals and food products manufacturing processes and quality control. It ensures that the products are safe, effective, and high-quality, protecting public health.

2. **Why validation of a manufacturing process is important?**
   Answer: The manufacturing process should be validated to ensure that the process consistently produces a product that meets the desired quality standards.

3. **Why are instruments calibrated?**
   Answer: Manufacturing equipment should be regularly calibrated and maintained to ensure it functions correctly and produces the desired results.

4. **Which stage of a drug lifecycle is GMP applied?**
   Answer: GMP applies to all drug product lifecycle stages: from the manufacture of drugs to technology transfer, commercial manufacturing, distribution, and discontinuation of the product.

5. What is the role of the Quality Department in production?
Answer: Sampling, testing, documentation and release.

6. What are the key components of Good Documentation practice?
Answer: Documentation should be clear, concise, accurate, and legible. Data entries should be recorded promptly when actions are performed. Backdating and postdating are not allowed. All corrections to the original entries should be initialled and dated. All the corrections must be properly explained.

7. What is the role of the Quality Department in production?
Answer: Sampling, testing, documentation and release.

8. How long documents should be retained?
Answer: It should be at least one year after the batch expiration date.

9. What should a training document contains?
Answer: The training documentation should include a training description including the name of training, version, mode (self-training or instructor-led), completion date and information on the trainer.

10. What should an analytical method contain?
Answer: An analytical procedure contains the purpose, test information, product information, materials and equipment, procedure, as applicable, System suitability requirements, solutions and reagents preparation, standards and samples preparation, instrument parameters, calculations, review and approval with approval dates and revision history.

11. How to store ram materials and finished products?
Answer: Raw materials and finished products should be stored, handled, and transported in a way that prevents contamination or damage

12. Why is Change Control required?
Answer: Change control is important for the following reason

1. To assure that changes do not impact Product Identity, Strength, Quality, Purity, Potency
2. Does not affect validated Processes/Systems
3. No impact on Regulatory Commitments (e.g., NDA / ANDA / BLA)
4. To provide a single, easy to access the source of information regarding changes for

    - Audits/Inspections
    - Annual product reviews (internal)
    - Annual reports (to FDA)

13. What do companies achieve by adhering to Change Control procedure?
Answer: By adhering to the Change Control policy and procedure, companies achieve the following:

1. Manufacture a product that meets the specification
2. Keep equipment free from failure
3. Ensure sterility assurance
4. Comply with cGMP requirements
5. Prevent product quality problems including FDA 483 findings, warning letters, sub-standard products, or product recall.

2 Concept of Calibration

Calibration is a process that compares the measurement values of measuring equipment against a reference standard and certifies the measurement accuracy. It is system performance checking.

For example, if a mixer is used to mix constituents of a tablet and the mixer is set up for 50 rounds per minute, the mixer has to be checked if it is mixing 50 rounds per minute. This check is known as calibration. In the pharmaceutical industry, all the equipment are calibrated regularly at a specific time to ensure that it performs accordingly.

Validation is a detailed, documented process of confirming that the equipment or machine is installed correctly, operating effectively, and performing without error. It is a performance analysis methodology.

Suppose the same mixture has some constituents. When a product is developed, the mixing is validated. At the time of process validation, mixing is started. After certain intervals, sampling is done and tested to verify the optimum time to mix that mixture when all the constituents are evenly distributed in the entire mixture.

Method validation is the process used to confirm that the analytical method for a specific test is suitable for its intended use. Results from method validation can be used for quality, reliability, and consistency of analytical results.

Calibration means system performance checking. At the same time, validation is a performance analysis methodology.

Calibration is an operation to confirm that the equipment is performing accurately. In contrast, validation is an operation to verify that the analytical method is performing accurately.

Why calibration and validation are essential?

For example, a coffee shop you regularly visit for your favorite tea. During a visit, you notice that the taste of tea is not as good as it used to be.

You enquire to the shop owner and may get one of the following reasons.

- They switch tea bags to open tea leaves or vice versa.
- They changed the equipment they used when boiling water. Before, they used electric heaters and now gas stoves.
- They changed their vendor for the milk supply.
- The chef who used to make tea has retired, and they hired a new chef.

What action will you take? You will most likely stop going to this place and look for another coffee shop.

Regarding medications, customers can't afford to do the same thing. Medications are critical for people's health and vital when saving someone's life. You need medication which is as effective as it is to be.

DOI: 10.1201/9781003366393-2

That's why in the pharmaceutical industry, each machine is calibrated, and each process is validated to ensure the production of medicine which constantly meets the specification.

If any machine is changed, the process at that machine is re-validated, ensuring that the process will not affect the quality of the product. Similarly, if the source of any raw material is changed, the process is re-validated.

In this chapter, the calibration of some common instruments will be discussed. Validation is discussed in detail in the method validation and process validation chapters.

CALIBRATION OF A FRIABILATOR

A friabilator is used to calculate the friability of tablets. In this instrument, a wheel is rotated 25 times per minute.

Turn on the instrument.

Use a calibrated stopwatch to measure the number of revolutions.

Using a calibrated stopwatch, count the number of revolutions the wheel completes in 1 minute.

Record the readings at three different time points. Calculate the mean.

The result should be 25 ± 1 round per minute.

CALIBRATION OF AN ANALYTICAL BALANCE

Check the balance manual for specific instructions on how to calibrate it. This will vary depending on the model of the balance.

Make sure that the balance is clean and free of debris. Use a clean, soft brush to remove any dust or particles.

Carefully wipe the pan using water first, then isopropanol, and a soft cloth. Allow to air dry. Brush out any dust and particles from the pan holder and surrounding area. Reposition the pan. Close the doors and allow the balance to dry for 10 minutes.

Turn on the balance and let it warm for at least 30 minutes.

Level the balance by adjusting the left and/or right legs so that the internal levelling indicator bubble falls within the back circle when viewed directly from above.

Tare the balance by pressing the "tare" key. Allow the balance to stabilize. The reading must be "0.0000". If the display does not read "0.0000", continue taring until the desired stable reading is achieved.

Select a calibration weight on balance. The weight should be the same as or slightly larger than the maximum capacity of the balance.

Place a certified weight for the balance in the middle of the pan and record the weight. Place weights in five different positions close to the edge of the pan and record the weights.

Wait for the balance to stabilize and display the weight of the calibration weight.

Record the weight displayed on balance and compare it to the known weight of the calibration weight.

If the weight displayed on balance is within the manufacturer's specified tolerance range, the balance is calibrated and ready to use. If the weight is outside the range, recalibrate the balance.

Repeat the calibration process with additional calibration weights if necessary.

Keep a record of the calibration process and results, including the date, weights used, and any adjustments made.

CALIBRATION OF AN HPLC COLUMN HEATER

Calibrating an HPLC column heater involves checking its accuracy and making necessary adjustments to ensure the correct temperature readings. Here are the general steps to follow:

Prepare the calibration equipment: You will need a temperature probe and a certified/calibrated thermometer.

Set the column heater to a temperature with the highest range of your calibration equipment.

Insert the probe of a calibrated digital thermometer through the frontal/lateral opening of the heater. Ensure the tip is in contact with the heat exchanger's floor.

Wait 30 minutes. Read the digital thermometer.

Check the temperature reading on the column heater against the temperature reading on the thermometer. If there is a significant difference, note it down.

Calculate the difference between the temperature reading on the column heater.

Adjust the column heater temperature accordingly, using the calculated difference as a guide. For example, if the column heater reads 5°C higher than the probe, lower the column heater temperature by 5°C.

Repeat the process several times, adjusting the temperature as necessary until the column heater temperature reading matches the temperature reading from the probe in the water.

Once you have completed the calibration process, record the adjusted temperature settings and keep them for future reference.

Repeat the process at the lower range of your calibration equipment.

It's important to note that the specific calibration process may vary slightly depending on the type and model of the HPLC column heater you are using. Be sure to consult the manufacturer's instructions.

CALIBRATION OF AN HPLC PUMP

Calibrating an HPLC pump involves checking the accuracy and precision of the flow rate. Here are the steps to follow:

Prepare the calibration solution: prepare a solution of known concentration suitable for the detection system used in the HPLC. Ensure the solution is stable and homogeneous.

Flush the system: Flush the HPLC system with the mobile phase solvent to remove contaminants or residue from previous use.

Install a suitable column: Equilibrate a right column with the mobile phase solvent to remove any air bubbles, then connect it to the HPLC pump.

Connect a restrictor loop to the pump's outlet solvent line.

Program a flow rate of 0.5 mL/min (or minimum desired flow). Allow the pressure to stabilize.

Obtain a 5 mL volumetric flask and a calibrated stopwatch. Insert the outlet from the restrictor loop into the volumetric flask and start the stopwatch simultaneously.

Stop the stopwatch when the bottom of the meniscus reaches the 5 mL mark on the volumetric flask. Record the time. (Time should be 10 minutes)

Repeat steps using flow rates of 2.0 and 5.0 mL/min (or the highest desired flow) and volumetric flasks of 10 and 25 mL, respectively. Record the results. (Time should be 5 minutes for both)

Calculate the accuracy and precision: Compare the measured flow rate with the set flow rate and calculate the accuracy and precision of the HPLC pump.

Adjust the pump: If the flow rate is not accurate, adjust the pump flow rate using the appropriate control knob or software.

Recheck the flow rate: After adjusting the pump, recheck it to ensure it is accurate and precise.

CALIBRATION OF AN HPLC DETECTOR

Calibrating an HPLC detector is essential in ensuring accurate and reliable results in chromatographic analysis. Here are the general steps to calibrate an HPLC detector:

Choose a suitable calibration standard: The standard should be stable, pure, and available in a high concentration. Typical standards include caffeine, toluene, and acetone.

Prepare the standard solution: Dissolve the standard in a suitable solvent to make a stock solution. Dilute the stock solution to prepare a series of standard solutions with different concentrations.

Inject the standard solutions into the HPLC: Inject each common solution into the HPLC system and record the detector response for each solution.

Plot a calibration curve: Plot a graph of detector response versus concentration for the standard solutions. Use a suitable regression analysis method to fit a straight line to the data points.

Verify the calibration curve's linearity: Check the calibration curve's linearity by injecting additional standard solutions at different concentrations and comparing the actual detector response with the response predicted by using the calibration curve.

Calculate the detector sensitivity: Determine the slope of the calibration curve, which represents the detector sensitivity. The sensitivity should be consistent across the range of concentrations tested.

Validate the calibration: Validate the calibration by injecting a sample of the analyte of interest and comparing the detector response with the response predicted by using the calibration curve.

CALIBRATION OF AN HPLC AUTOINJECTOR

Calibrating an HPLC autoinjector is essential in ensuring accurate and reliable results in chromatographic analysis. Here are the general steps to calibrate an HPLC autoinjector.

Calibrating an HPLC autoinjector involves several steps to ensure that samples are accurately and precisely injected. Here are the general steps for calibrating an HPLC autoinjector:

Set up the HPLC system: Ensure that the HPLC system is properly set up and is in good working condition. Check that the system is clean and free of any debris or contaminants.

Check the sample loop: Verify that the sample loop is properly installed and that there are no leaks or blockages in the loop.

Prepare the HPLC system for analysis. Install a column and prepare standard and mobile phase. Condition the HPLC system with the mobile phase.

Program the system to perform six injections per selected volume of the Autoinjector.

Vial 1: 10 µL, Vial 2: 50 µL, Vial 3: 100 µL, and Vial 4: 200 µL.

Precision: calculate the peak area %RSD for each set of six injections.

Limits: % RSD NMT 1% (Typical value 0.75%)

Linearity: calculate the average peak area for each set of six injections. Perform a linear regression using the nominal injection values (Y values) and the average peak areas (as X values). Calculate the regression coefficient (r).

Limits: r NLT 0.999

CALIBRATION OF DISSOLUTION APPARATUS

Physical Check

Calibrating a dissolution unit involves verifying and adjusting its performance to ensure accurate and consistent results.

Inspect baskets, paddles, shafts, and vessels for cleanliness. Inspect baskets, basket-shaft and paddle-shaft assemblies, and vessels for deformation, burrs, or other defects (defects include misshaped or tom baskets, gouged paddles, Teflon coating peeling off, and goss scratches.

The centre position of the paddle and basket shafts should be analyzed. Clip the calibrated centering gauge onto the shaft and allow it to come into contact with the vessel wall. Slowly rotate the gauge, determine the maximum deviation from the center, and record each shaft. It should be within 2 mm from the center.

Verify the distance between the vessel bottom and each shaft. (Paddle and Baskets) using a calibrated height check gauge. This height must be 23–27 mm.

Using an appropriate levelling device (e.g., bubble level), ensure that the bath is level at all four corners. Make adjustments if necessary.

Measure the shaft revolutions per minute by means of a calibrated portable tachometer at 50, 100, and 200 RPM. It should be within 4%.

Verify the temperature sensor: Check the temperature sensor's accuracy by immersing it in a temperature-controlled water bath and comparing the readings with the actual temperature of the bath. Adjust if necessary.

Verify the volume: Measure the volume of the dissolution vessel and ensure that it matches the manufacturer's specifications. Adjust if necessary.

Chemical Check

USP Prednisone Tablets, 10 mg, are sued for calibration. Degas 10 L purified water.

Prepare a 15 µg/mL solution of Prednisone USP standard in water. Add 500 mL degas water to six dissolution vessels. Set up the temperature of the dissolution apparatus at 37°C and rotation at 50 RPM. Check the water temperature, equilibrate the temperature at 37°C, stop the rotation, add one tablet to each vessel, and start the rotation and the timer for 30 minutes.

Withdraw about 20 mL of solution. Using a spectrophotometer, analyze samples using standard at 242 nm. Refer to the certificate of analysis of the USP tablet used for specification.

CALIBRATION OF A PH METER

pH is a critical test for in-process and finished products.

Usually, it is tested for liquids and creams, but in some sold dosages, it is also performed by preparation of aqueous solutions of solids.

pH is the inverse log of the hydrogen ion concentration in a liquid. pH is very important in the process and finished product tests for liquid medicine. It determines how acidic or basic a solution is.

The instrument used to measure the pH is known as the pH meter. The pH meter is commonly used in laboratories.

The pH meter consists of a pH sensor and pH signals can be transmitted digitally to an external device such as a computer or any other display.

pH Measurement System

The measurement system consists of three units:

1. **Measuring Electrode**

   It is sensitive to hydrogen-ion activity, typically a glass electrode, though other electrode types are possible.

2. **A Suitable Reference Electrode**

   A reference electrode is used.

3. **A Voltage Measurement System**

   A voltage measuring system with an input resistance capable of measuring at a high input impedance of the pH sensor.

The measuring and reference electrodes may be separated or combined. The voltage measurement system may be separated from the pH sensor or integrated into the sensor. For most applications, a temperature measurement will be necessary. A temperature device may be embedded into the pH sensor, or an external temperature device may be used.

Buffer Solutions for Calibration of the pH Measurement System

Buffer solutions for calibration can be prepared. Buffer salts of requisite purity can be obtained from commercial suppliers. Buffer solutions should be stored in appropriate containers that ensure the stability of the pH through the expiry date and fitted with tight closure.

Electrode Preparation and Maintenance

Electrodes must be rinsed with distilled water and blotted, not rubbed dry, with lint-free tissues. Electrodes must not be exposed to air for more than 30 seconds.

Use internal filling solution Ag/AgCl

While not in use, electrodes are stored 3.5M KCl/AgCl solution in water.

Calibration

Several pH meters are available in the market, and their calibration procedures me be different. General principles can be followed.

Commercially available buffer solutions for pH measurement systems may be used. Solutions prepared from ACS reagent grade materials or other suitable materials may be used, provided that the pH of the resultant solution is the same as that of the solution prepared from the NIST (or other national authorities) certified material. Buffer solutions over 12 should be used immediately or prepared using freshly boiled water and stored under conditions to minimize carbon dioxide absorption and ingress.

Rinse the electrode first with DI water and then with each specific buffer before placing it in the solution.

A two-point calibration can be performed. Three buffer solutions are to be used for calibration and verification. Two of the buffer solutions are used for the calibration process, with the pH of the testing material expected to fall within this range. The third buffer solution with intermediate pH is used to verify calibration accuracy.

The pH range between the two selected calibration buffer solutions should be kept narrow to ensure better accuracy of the pH reading. See the below examples:

Calibration using pH 1.00 buffer and pH 4.00 buffer, pH 1.68, or 2.0 buffer can be used for verification. Ideally, two buffers should not be used with a difference of 4 pH.

Calibration must be done such that the target pH is within the range.

Calibration using pH 4.00 buffer and pH 7.00 buffer, pH 6.86 buffer can be used for verification.

Calibration using pH 7.00 buffer and pH 10.00 buffer, pH 9.18 buffer can be used for verification.

Calibration using pH 9.18 buffer and pH 12.00 buffer, pH 10.00 buffer can be used for verification.

Procedure for Calibration

Turn on the Instrument

Before calibrating and using your pH meter, you must turn it on and allow adequate time for the meter to warm up. This should generally take around 30 minutes but check your pH meter's operating manual for exact times.

Change the mode of pH meter to calibration. Immerse the pH probe in the first buffer, wait to stabilize the reading, and press the calibration button.

Now rinse the pH probe with purified water, dry it, and immerse it in a second buffer; wait to stabilize the reading and press the calibration button when the pH display becomes constant. Record the slope and offset value. Now change the mode to pH reading. Read the checking buffer; the pH of the checking buffer must be within ± 0.05 pH.

Rinse your electrode with distilled water and blot or dry with a lint-free tissue.

Discard the buffer when you are finished. Do not return it to its original container.

After completing the two-point calibration process (before measuring the pH of the verification solution), verify that the pH slope and offset are within acceptable parameters.

Typical acceptable parameters are a 90%–105% slope and a 0 ± 30 mV offset. Depending on the pH instrumentation, the pH slope and offset may be determined in software or by manual methods. If using manual methods, follow supplier instructions to calculate the pH sensor slope/offset. If these parameters are not within acceptable parameters, the sensor should be cleaned appropriately, serviced, or replaced, and the two-point calibration process shall be repeated.

Now the pH meter is calibrated and ready for use.

Prepare the test material according to the monograph's requirements or specific procedures. If the pH of the test sample is sensitive to ambient carbon dioxide, then use purified water that has been recently boiled and subsequently stored in a container designed to minimize the ingress of carbon dioxide.

Rinse the pH sensor with water, then with a few portions of the test material.

Immerse the pH sensor into the test material and record the pH value and temperature.

In all pH measurements, allow sufficient time for stabilization of the temperature and pH measurement.

VALIDATION

The details about method validation and process validation are included in other chapters.

QUESTIONS AND ANSWERS

1. **Why are calibration and validation necessary in the pharmaceutical industry?**
   Answer: To ensure the production of medicine which constantly meets the specification.

2. **What is the difference between calibration and validation?**
   Answer: Calibration is an operation to confirm that the equipment is performing accurately. In contrast, validation is an operation to verify that the analytical method is performing accurately.

3. **How the range of HPLC pump is selected for calibration? Give an example.**
   Answer: HPLC pump is calibrated in the range of its use. Suppose a pump is used from 0.5 to 5.0 mL/min. The pump will be calibrated at 0.5–5.0 mL/min. The instrument cannot be used outside this range.

4. **Which tablet is used to calibrate the dissolution apparatus?**
   Answer: USP Prednisone Tablets, 10 mg, are sued for calibration.

5. **What should be the distance between the dissolution basket/paddle and the bottom of the vessel?**
   Answer: 23–27 mm

6. **What should be the temperature of the dissolution bath?**
   Answer: 37°C

7. **What is the rotation of speed of a friabilator?**
   Answer: 25 RPM

8. **If the pH of a solution needs to be adjusted, which range of calibration will be used and what buffer will be used for pH check?**
   Answer: Range 1.00–4.00, pH check with pH 1.68.

9. **What should be a reasonable difference between the value of two buffers when calibration is done?**
   Answer: Maximum 4 unit difference.

10. **What are the acceptance value of slope and offset while calibrating a pH meter?**
    Answer: Typical acceptable parameters are a 90%–105% slope and a 0 ± 30 mV offset.

11. **What solution is used to store the pH electrode when a pH meter is not in use?**
    Answer: While not in use, electrodes are stored 3.5M KCl/AgCl solution in water.

3 Reference Standards and Reagents

REFERENCE STANDARDS

A reference standard is a specific batch of a substance of known composition and/or purity to which the responses of samples of unknown composition or purity are compared to determine the concentration or purity of that unknown composition.

In other words, an ultra-pure compound of a known potency is used for quantitative analysis. The solution of this known compound and analyte is analyzed. The response of the analyte is compared with this ultrapure compound, and the concentration of the analyte is calculated. This ultra-pure compound is known as "Standard".

PRIMARY STANDARDS

USP or other compendia can purchase the standard; their potencies are already established by compendia. These standards are known as primary standards.

SECONDARY STANDARDS

Raw material can also be used if qualified using a USP standard. This qualified standard is known as the in-house standard or secondary standard.

USP STANDARDS

United States Pharmacopeial provides the standard known as USP Reference Standards (USP RSs). USP standards are highly characterized materials demonstrated to have the appropriate qualities to support their intended use. USP RS is not for use in humans or animals.

USP RS is used for tests including in the United States Pharmacopeia (USP) or National Formulary (NF) documentary standards.

Those standards have been approved and established as suitable for use in the context of these tests.

USP RS may also be used in those tests which USP does not support. Assessment of the suitability for use in other applications is the responsibility of the user.

The value assigned to quantitative standards depends upon a mass determination by a primary reference measurement procedure such as mass balance determination. If the value has been assigned using a primary reference measurement procedure, the USP RS in metrology terms can be considered a primary measurement standard.

DOI: 10.1201/9781003366393-3

If the value has been assigned by comparison or calibration to another material, the USP RS should be considered a secondary measurement.

The material and performance attributes for these USP RS are described in the supporting documentation supplied with the USP RS.

USP RS goes through extensive characterization as part of its establishment process. The official uses of the standard primarily drive the types and extent of testing, but material characterization goes beyond the establishment of suitability for use. Typically, a comparison to the previous lot is performed during the study when establishing a replacement lot of a USP RS to verify the suitability of the new standard lot.

USP or NF monographs and general chapters specify official applications of USP RS. These applications are as follows. Some USP RSs could be used for several types of applications listed below.

QUANTITATIVE DETERMINATIONS

The majority of USP RSs for quantitative determinations support measurements for total amounts of material on a mass basis. This category includes USP RS for USP or NF articles and impurity standards labelled for quantitative use. The assigned value of the USP RS is stated on the labelling and should be included in calculations used in the monograph and applicable general chapters.

USP RS for relative determinations of potency or activity is often required for the measurement of complex materials and quantitative amounts may be expressed in units or relative potency terms other than mass.

These USP RSs are established by calibration to a primary standard where the property of the material determines the unit. The USP RS documentation will indicate when the USP RS has been established by comparison to an International Standard (IS) established by the WHO. Additional statements about unit/mass relationships, specific activity, or other relevant information related to the measurement may be provided in the USP RS documentation.

For antibiotics that use microbial assays to determine activity, the potency is determined in units or micrograms per milligram (µg/mg) of activity. The units or µg/mg of activity are established against a WHO IS when one exists for that antibiotic. Where no WHO IS presently available, USP establishes and maintains the standard to calibrate USP RS lots.

QUALITATIVE DETERMINATIONS

Identification USP RS: USP RS for identification tests are typically presented as single components of high chemical purity but may also be complex materials of natural, synthetic, or recombinant origin (e.g., biologics, natural products, botanicals, complex non-biological products, and others). For complex materials, the identification attributes are presented in a matrix of other materials and require highly specific measurement systems, such as nucleic acid-based identity determination for naturally derived materials.

IMPURITY USP RS

Impurity USP RS is typically used for system suitability or as an impurity. They may be presented as single-component materials, as mixtures containing more than one impurity, or as drug substance(s) containing one or more impurities.

DIGITAL AND VISUAL USP RS

Unlike chemical reference materials, these USP RSs are not physical materials used in chemical analyses. Instead, these visual images are used by analysts to compare test articles to ensure that they meet compendial requirements.

PERFORMANCE VERIFICATION

These USP RSs are typically called for in general tests and assays and are provided to analyze and, where appropriate, to facilitate adjustment of the operation of an instrument to ensure the results obtained are accurate and/or precise or otherwise give acceptable results.

LABELLING

The labelling material contains the label to the USP RS and the associated USP Certificate. Both must be reviewed prior to handling or using the USP RS because, in some cases, not all of the necessary information can fit on the affixed label. USP Certificates are very specific.

The USP RS label typically contains the RS name, catalogue number, lot number, package size, assigned value, storage conditions, handling instructions, and country of origin information. For multi-component items, there is also an outer package and label.

The label also includes hazard and precautionary statements required by the Occupational Safety and Health Administration (OSHA).

In addition to the information provided on the USP RS label, the USP Certificate will generally contain the RS chemical name and structure, CAS number, molecular formula, and molecular weight. A typical chromatogram may also be included if necessary for the intended use. Additional information may be included, such as special handling instructions or information needed for using the USP RS. The USP Certificate also includes a copy of the label text and a series of general instructions.

PACKAGING

The amount of material per container depends on the application of the standard. Some standards are provided in single-use containers. Some single-use products may be lyophilized with content labelled in mass or activity units per container. If so labeled, the content of the container must be reconstituted in its entirety without any additional weighing. Instructions for use are given either on the label or USP Certificate or in the monographs where the standard is used.

STORAGE

USP RS should be stored in the packaging configuration provided by USP, according to the label and USP Certificate instructions. When storage in a refrigerator or freezer is stated on the label, follow the Packaging and Storage Requirements definitions. If no specific directions or limitations are provided on the label, the conditions of storage shall be room temperature and protection from moisture, light, freezing, and excessive heat.

Any unused portions remaining after the container has been opened should be carefully stored in accordance with the user's Standard Operating Procedures and good laboratory practices.

SUITABILITY FOR USE

USP periodically re-evaluates all USP RS throughout their lifecycles. The USP Continued Suitability for Use (CSU) program monitors real-time suitability for using all current lots of USP RS. Suitability testing intervals are established based on collaborative study data, manufacturer or supplier data, testing results, and CSU data trending and projections.

VALID USE DATE

USP RS lots are assigned a valid use date. The valid use date is the last day upon which a particular lot of USP RS can be used. Typically, the valid use date assigned is 1 year from the date the last vial of a lot is sold.

PROPER USE

Whenever the labelled directions for use require either drying or a correction for water and/or volatiles, this should be performed at the time of use.

The following list of label and definitions is provided as guidance for the handling and use of USP RS:

ASSIGNED VALUE (CALCULATION VALUE)

The quantity value assigned to a USP RS for its use in the quantitative compendial applications.

AS IS

Use the USP RS as received, without drying or additional testing and apply the assigned value to correct the concentration of the standard solution and/or preparation.

For the USP RS to be used on the as-is basis, the assigned value has already been corrected for volatiles, including moisture.

ANHYDROUS BASIS

Determine Water Content Titrimetrically at the Time of Use: Use the USP RS as received, and apply the water content determined to correct the weight of the standard. After the correction for water is applied, use the assigned value to correct the concentration of the standard solution and/or preparation.

At the time a USP RS is to be weighed, proceed with Water Determination and determine the water content on a separate portion of material. Instrumental or microanalytical methods are acceptable for this purpose.

DRIED BASIS

Determine Loss on Drying at Time of Use: Use the USP RS as received and apply the loss on drying value obtained to correct the weight of the standard. After the correction for loss on drying is applied, use the assigned value to correct the concentration of the standard solution and/or preparation. Determine the loss on drying value on a separate portion of the material, following the monograph procedure under Loss on Drying.

DRIED MATERIAL

Dry the USP RS before use. Use immediately after drying under stated conditions. Drying should not be performed in the original container. A portion of the material should be transferred to a separate drying vessel.

REAGENTS

Some common reagents are required in the pharmaceutical laboratory for all tests. Some classes of reagents will be discussed in this chapter.

ACS (American Chemical Society) Reagent Grade

When the method indicates to use of "ACS reagent grade", it is intended that a grade meeting the corresponding specifications of the current edition of Reagent Chemicals shall be used.

Suitable Grade

If an ACS grade reagent is unavailable, then a suitable reagent grade commercially available shall be used.

Packaging and Storage

Reagents and solutions should be preserved in tight containers of resistant glass or other suitable material. Directions for storage in light-resistant containers should be carefully observed.

Stoppers and stopcocks brought into contact with substances capable of attacking or penetrating their surfaces may be given a protective coating of a thin film of a suitable lubricant.

Metal-Ion Standard Solutions

Atomic absorption and flame photometry require the use of a number of metal-ion standard solutions. While the individual monographs usually provide directions for the preparation of these solutions, the use of commercially prepared standardized solutions of the appropriate ions is permissible, provided that the analyst confirms the suitability of the solutions and has data to support their use.

Some Definitions

Reagent Test Solution

Reagents are substances used either as such or as constituents of solutions.

Dry Chemicals/Liquid Chemicals

Dry chemicals/liquid chemicals are substances used either as such or as constituents of solutions.

Indicators

Indicators are reagents used to determine the specified endpoint in a chemical reaction, to measure hydrogen-ion concentration (pH), or to indicate that a desired change in pH has been affected.

Buffer Solutions

Buffer solutions resist changes in the activity of an ion with the addition of substances that are expected to change the activity of that ion. pH buffer solutions are labelled with a pH value accurate to 0.01 pH units. Buffer solutions are used for the standardization of pH meters.

Colorimetric Solutions (CS)

Colorimetric solutions are solutions used in the preparation of colorimetric standards for comparison purposes.

Test Solutions (TS)

Test solutions are solutions of reagents in such solvents and of such definite concentrations as to be suitable for the specified purposes.

Volumetric Solutions (VS)

Volumetric solutions are solutions of reagents of known concentrations intended primarily for use in quantitative determinations. Concentrations are usually expressed in terms of normality (N) or molarity (M).

QUESTIONS AND ANSWERS

1. **What is a reference standard?**
   Answer: The reference standard is an ultra-pure compound of a known potency used to analyze by comparing its response with the analyte response.

2. **What is the difference between primary and secondary standards?**
   Answer: Primary standard is purchased by USP or other compendia that have already established their potencies, while the secondary standard is a raw material which is qualified against the standard from USP or other compendia.

3. **Can a USP standard be used for a test not supported by USP?**
   Answer: Yes, a USP standard can be used for a test not supported by USP.

4. **What is an Impurity USP RS?**
   Answer: They may be presented as single-component materials, as mixtures containing more than one impurity, or as drug substance(s) containing one or more impurities. Impurity USP RS is typically used for system suitability or as an impurity.

5. **How is the standard used on as dry basis?**
   Answer: "Loss on drying" is determined or the standard and this value of LOD is used as standard potency. Loss on drying value is determined on a separate portion of the material.

6. **How is an indicator?**
   Answer: Indicators are reagents used to determine the specified endpoint in a chemical reaction, to measure hydrogen-ion concentration (pH), or to indicate that a desired change in pH has been affected.

7. **What is a buffer solution?**
   Answer: Buffer solutions resist changes in the activity of an ion with the addition of substances that are expected to change the activity of that ion. Buffer solutions are used for the standardization of pH meters.

8. **What is a volumetric solution?**
   Answer: Volumetric solutions are solutions of reagents of known concentrations intended primarily for use in quantitative determinations. Concentrations are usually expressed in normality (N) or molarity (M).

9. **How is a dried standard used?**
   Answer: Dry the Standard before use. Use immediately after drying under stated conditions. Drying should not be performed in the original container. A portion of the material should be transferred to a separate drying vessel.

10. How should reagents and solutions be stored?

Answer: Reagents and solutions should be preserved in tight containers of resistant glass or other suitable material. Directions for storage in light-resistant containers should be carefully observed.

Stoppers and stopcocks brought into contact with substances capable of attacking or penetrating their surfaces may be given a protective coating of a thin film of a suitable lubricant.

4 Chemical Tests

SOME COMMON TESTS AND CALCULATIONS

Several tests are performed in the Quality control chemistry laboratory to analyze the raw and finished product composition. The reason to carry out these tests is regulatory. One must know all about the particular batch of the material for its composition, potency and impurities before using it to manufacture a finished product, whether it is an excipient or an active pharmaceutical ingredient. Therefore, the Quality control chemistry laboratory performs specific tests to analyze all such materials. The most common tests which all active materials pass through are assay content, impurities and related substances. However, there are still many more.

In this chapter, these tests and some other common tests will be discussed.

In Chapter 9 of this book, the instruments involved in these tests and their usage details will be covered.

A validated method is always used in quality control to test a specific product. Analysts must follow the method strictly. If any deviation is done from the method, it has to be documented and authorized.

ASSAY AND CONTENT UNIFORMITY

The assay is an analytical procedure determining the presence of an active pharmaceutical ingredient in a dosage form.

All dosages (Tablets, capsules, cream, ointment, liquid and sprays) contain some active pharmaceutical ingredients and excipients.

Non-medical ingredients of any medicine are called excipients.

For example, we see a tablet in different potencies, such as 50 or 100 mg.

50 mg tablet may weigh 200 or 300 mg, but other than 50 mg active, all material is excipients.

The role of excipients is to dissolve the tablet in a certain condition and in certain parts of the body.

An assay test is done on a representative sample containing a composite sample of multiple dosages combined and tested.

In Content uniformity, ten or more dosages are tested individually to determine the content of active in a single dose.

AN EXAMPLE METHOD FOR ASSAY

Assay by HPLC

Mobile phase: Methanol: Water (75:25)
Diluting solvent: Methanol

 DOI: 10.1201/9781003366393-4

HPLC Parameter

Injection volume: 50 μL

Wave length: 240 nm
Run time: 10 minutes
Flow rate: 1.0 mL/min

Standard Stock Preparation

Accurately weigh 100 mg standard and transfer into a 100 mL volumetric flask. Dissolve and dilute to volume with methanol.

Standard Working Preparation

Pipette 5.0 mL standard stock to a 100 mL volumetric flask and dilute to volume with Methanol.

Sample Stock Preparation

Transfer 10 whole tablets to a 1000 mL volumetric flask. Add 500 mL methanol, sonicate for 10 minutes and shake on a shaker for 20 minutes. Dilute to volume with methanol, mix well and filter.

Two types of samples are usually prepared for assay.

Procedure

Inject standard working solution five times, RSD of the area of five injections should be not more than 2%. Inject sample working solution. Inject a standard working solution in the end.

Calculations

% Assay = Area of sample/Area of standard x weight of standard x dilution of sample/dilution of standard × 1/number of tablets × 100/label claim

Whole Tablet/Capsule

A specific number of dosages are taken to prepare the sample.

Composite Sample Preparation

A specific number of dosages are combined, and a portion is weighed.

The sample is analyzed using a standard solution.

Example for Whole Tablet Analysis

10 tablets of 50 mg each dissolved in 1000 mL diluent.

50.23 mg standard dissolved in 100 mL (Standard potency 99.8%

5.0 mL of resulting solution diluted to 100 mL.

5.0 mL of resulting solution diluted to 100 mL.

Standard and sample analyzed using HPLC

The average area from the standard was found to be 3,025,464
The sample area was found to be 2,987,846
Calculation:

% Assay = Area of sample/Area of standard × weight of standard × dilution of sample/dilution of standard × 1/number of tablets × 100/label claim

Corrected weight for standard: 50.23 × 99.8/100 = 50.13 mg
Dilution of sample = 1000 × 100/5 = 20,000
Dilution of Standard = 100 × 100/5 = 2000
Label claim = 50 mg/tablet
% Assay = 2,987,846/3,025,464 × 50.13 × 20,000/2000 × 1/10 × 100/50 = 99.0%

Example for Composite Tablet Powder Analysis

In this procedure, a specific number of tablets are grounded, and power is weighed for the analysis.

20 tablets of 25 mg each are weighed to determine the average tablet weight.

Tablets are grounded to small particles.

The weight of 20 tablets is 2000.2 mg, so the average weight is 100.01 mg.

Tablet powder equivalent to 100 mg active is to be weighed. (It should be about 400 mg)

The powder weight is 400.2 mg.

This powder is dissolved in a 200 mL volumetric flask. 10.0 mL of resulted solution is diluted to 100 mL.

50.25 mg (The potency of the standard is 99.9%) standard is dissolved in a 100 mL volumetric flask. 10.0 mL of resulted solution is diluted to 100 mL.

Calculation:

% Assay = Area of sample/Area of standard × weight of standard × dilution of sample/dilution of standard × Average weight of the tablet/weight of tablets powder × 100/label claim

Corrected weight for standard: 50.25 × 99.9/100 = 50.20 mg
Dilution of sample = 200 × 100/10 = 2000
Dilution of Standard = 100 × 100/10 = 1000
Label claim = 25 mg/tablet
The average area of standard 201,546
The average area of standard 198,745
% Assay = 198,745/201,546 × 50.20 × 2000/1000 × 100.01/400.2 × 100/25 = 99.0%

Content Uniformity

The content uniformity test determines the presence of an active ingredient in a single dosage. It can be a single tablet, capsule or spray.

The test is similar to the whole tablet assay. The sample is prepared using one tablet, capsule or spray.

IMPURITIES OR RELATED SUBSTANCES

Impurities are critical quality attributes of drug substances and products because they can affect their safety and efficacy.

Impurity measurements for drug products can be challenging using chromatographic methods due to the lower amount of impurity in a product.

Most pharmacopeia does not quantify in drug products impurities that are evaluated in drug substances and that are not expected to increase in the drug product during manufacturing or storage (i.e., impurities that are not degradation products).

DRUG SUBSTANCE

Impurities in drug substances can be classified into the following categories:

Organic impurities
Inorganic impurities
Residual solvents

Organic impurities arise during the manufacturing process and/or storage of the drug substance. They can be identified or unidentified, volatile or non-volatile substances. It includes the following:

Starting materials
Reaction by-products
Intermediates
Degradation products
Reagents and ligands

Inorganic impurities can result from the manufacturing process. They usually are known and identified and include the following:

Reagents and catalysts
Elemental impurities
Inorganic salts
Other materials

Elemental impurities can include catalysts and environmental contaminants. These impurities may occur naturally, be added intentionally, or be introduced inadvertently (e.g., by interactions with processing equipment and the container).

Residual solvents originate from organic liquids used as vehicles in synthesizing a drug substance, as starting material, or as the intermediates involved in the synthesis.

Documented evidence that the analytical procedure used to evaluate impurities is validated and suitable for detecting and quantifying impurities should be established.

DRUG PRODUCT

The specification for organic impurities in a drug product should include a list of degradation products expected to occur during the manufacture of the commercial product and under the recommended storage conditions. Stability studies, forced degradation studies, knowledge of degradation pathways, product development studies, compatibility studies, and laboratory studies should be used to characterize the degradation profile. The selection of degradation products in the drug product specification should be based on the degradation products found in batches manufactured by the proposed commercial process.

For drug products, the concept of setting degradation product limits is based on the sound scientific assessment as applied to available data on the safety and stability of the drug product, data that may include the degradation pathways of the drug substance, the manufacturing process, known excipient interactions, any safety assessment studies (e.g., predictive toxicology programs, mutagenicity assessments, and in vitro and in vivo toxicology studies), stability studies conducted under the recommended storage conditions, and ancillary studies that may provide additional information on the stability profile of the drug product. Impurities that are not degradation products (i.e., process-related impurities from the drug substance) are often not controlled in the drug product, as they are typically controlled in the drug substance and these impurities are not expected to increase over time. In cases where the studies show that impurities from the drug substance can increase over time (i.e., they are also drug product degradation products), they should be controlled in the drug substance and the drug product.

ORGANIC IMPURITIES IN DRUG SUBSTANCES AND DRUG PRODUCTS

The organic impurities to be controlled in drug substances are process-related impurities and degradation products. The organic impurities to be controlled in the drug product are those resulting from the drug substance's degradation or the drug substance's interaction with excipients and/or the primary container–closure system. Drug substance process-related impurities need not be controlled in the drug product unless they are also degradation products. However, in some cases, drug substance process-related impurities may be included in the drug product specifications, if appropriate, and limited by appropriate acceptance criteria.

Manufacturers are responsible for controlling organic impurities in accordance with current regulatory standards.

Quantification of Impurities

As compared to the assay test, in impurities quantification, the standard is prepared in lower concentration because the quantity of impurities is much lower and for precise results, standard and impurity should be close.

On the other hand, sample solutions are prepared in higher concentrations. If the concentration of the solution is low, the impurities are difficult to quantify.

There are two types of impurities are being quantified.

1. **The known impurity**

   Sometimes the standard of known impurity is prepared and known impurity is quantified with the response of the impurity standard.

   In some cases, the standard of impurity is not available, the impurity is quantified with the response of active standard. In this case, relative retention time should have been established at the time of method development.

2. **The unknown impurity**

   The unknown impurities are quantified based on active standard peak only.

An Example Method for Impurities

Impurities by HPLC

Mobile phase:	Methanol:Water (75:25)
Diluting solvent:	Methanol

HPLC Parameter

Injection volume:	50 μL
Wave length:	240 nm
Run time:	60 minutes
Flow rate:	1.0 mL/min

Standard Stock Preparation

Accurately weigh 5 mg standard and transfer into a 100 mL volumetric flask. Dissolve and dilute to volume with methanol.

Standard Working Preparation

Pipette 2.0 mL standard stock to a 100 mL volumetric flask and dilute to volume with Methanol.

Sample Working Preparation

Weigh 20 tablets and determine the average tablet weight. Grind tablets in fine powder. Weigh tablet powder equivalent to 100 mg of active. Transfer powder to a 100 mL volumetric flask.

Add 50 mL methanol, sonicate for 10 minutes and shake on a shaker for 30 minutes. Dilute to volume with methanol, mix well and filter.

Placebo Working Preparation

Weigh placebo powder equivalent to 100 mg of active. Transfer powder to a 100 mL volumetric flask.

Add 50 mL methanol, sonicate for 10 minutes and shake on a shaker for 30 minutes. Dilute to volume with methanol, mix well and filter.

Procedure

Inject standard working solution six times, RSD of the area of six injections should not be more than 2%. Inject blank (Diluting solvent), Placebo solution and sample working solution. Inject a standard working solution in the end.

Calculations

Ignore all sample peaks appearing in placebo. The limit of quantification is 0.05%, so do not report any peak which is less than 0.05%

% Impurity = Area of impurity in sample/Area of standard × weight of standard × dilution of sample/dilution of standard × Average weight of the tablet/ weight of tablets powder × 100/label claim

Example Calculation

In this procedure, a specific number of tablets are grounded, and power is weighed for the analysis.

20 tablets of 25 mg each are weighed to determine the average tablet weight.

Tablets are grounded to small particles.

The weight of 20 tablets is 2000.2 mg, so the average weight is 100.01 mg.

Tablet powder equivalent to 100 mg active is to be weighed. (It should be about 400 mg)

The powder weight is 400.2 mg.

This powder is dissolved in a 100 mL volumetric flask.

5.02 mg (The potency of the standard is 99.8%) standard is dissolved in a 100 ml volumetric flask. 2.0 mL of resulted solution is diluted to 100 mL.

% Impurity = Area of impurity in sample/Area of standard × weight of standard × dilution of sample/dilution of standard × Average weight of the tablet/ weight of tablets powder × 100/label claim

Corrected weight for standard: 5.02 × 99.8/100 = 5.010 mg

Dilution of sample = 100

Dilution of Standard = 100 × 100/2 = 5000

Label claim = 25 mg/tablet

The average area of standard 40,154

The area of impurity in sample 1987

Average Tablet weight = 100.01

Weight of sample = 400.2

% Impurity = 19,870/40,154 × 5.010 × 100/5000 × 100.01/400.2 × 100/25 = 0.050%

DISSOLUTION

The release of a drug from solid dosage in a certain medium and condition in the human body is called dissolution.

When a solid dosage like a tablet or capsule is taken orally, the medicine releases at a specific time and condition. This process is called dissolution.

Dissolution is different from disintegration. In disintegration tests, the solid dosage is broken into small pieces in a particular medium at a specific time. Disintegration does not determine how much drug will be released into the body.

The administration of drugs via oral dosage forms is one of the most common and effective means of delivering medicines to patients. When a dosage form is swallowed, the rate at which it releases the active ingredient is critical to ensure the drug is appropriately delivered. The rate at which the drug is released is called the dissolution rate.

Dissolution testing was initially developed for oral dosage forms, but the role of the test has now been extended to drug release studies on various other forms, such as topical and transdermal systems and suppositories.

Types of Dissolution Apparatus

Type 1:

Basket apparatus 50–120 rpm Conventional tablets, chewable tablets, CR (Controlled release)

Type 2:

Paddle apparatus 25–50 rpm orally Disintegrating tablets, chewable tablets, CR, suspensions

Type 3:

Reciprocating cylinder 6–35 rpm CR, chewable tablets

Type 4:

Flow through cell apparatus N/A ER, poorly soluble API, powder, granules, microparticles, implants

Type 5:

Paddle over disk 25–50 rpm Transdermal

Type 6:

Cylinder, Transdermal

Type 7:

Reciprocating holder 30 rpm CR (non-disintegrating oral and Transdermal)

The most common apparatus used are apparatus I (Basket) and apparatus II (Paddles). In this chapter, these two apparatuses will be discussed in detail.

Apparatus 1 – The Rotating Basket Dissolution Apparatus

The apparatus consists of a metallic drive shaft connected to the cylindrical basket. The basket is positioned inside a vessel made of glass or other inert, transparent material. The temperature of the media inside the vessel is kept constant by a water bath. The solution in the vessel is stirred smoothly by the rotating stirring element.

The diagram shows a simple graphic of the rotating basket setup. This is very tightly specified however and the official drawing according to the USP is shown below (Figure 4.1).

The drawing shows a basket manufactured from mesh, connected to a basket hub with a vent hole by using clips. The basket hub is connected to a rotating shaft. Every parameter of those individual parts is specified.

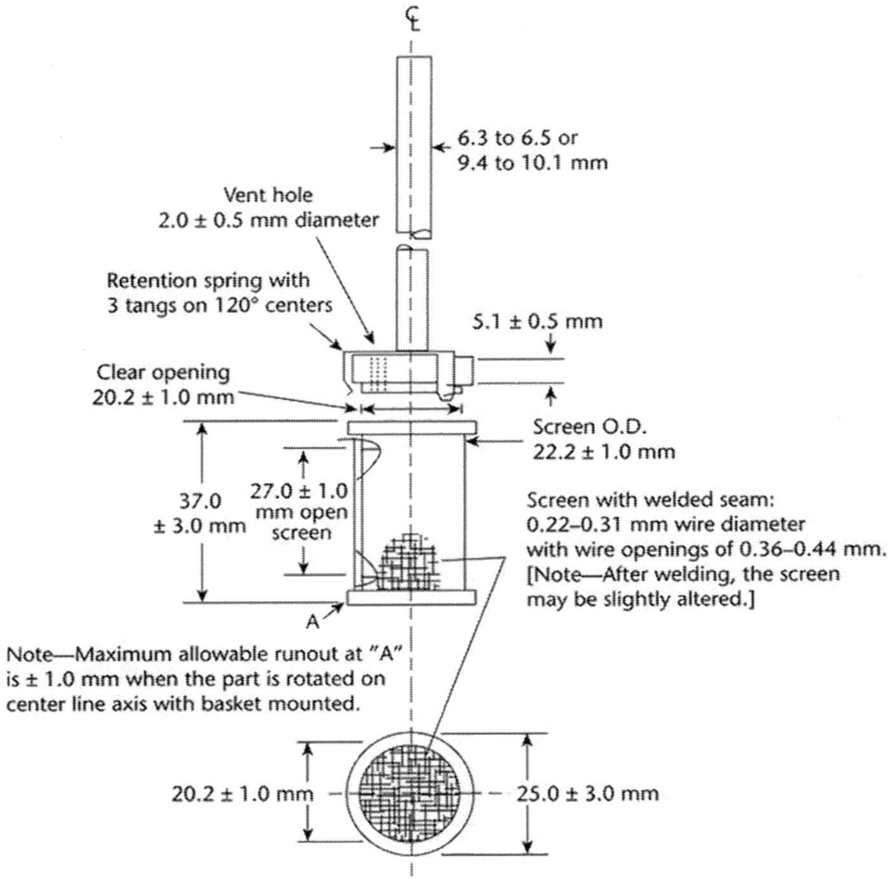

FIGURE 4.1 Dissolution apparatus I (basket).

The shafts and baskets must rotate freely and within specified limits, and importantly, they must be accurately centered in the vessel and be free from bubbles.

The standard basket consists of a stainless steel 40 mesh construction, but there are many variations depending on the monograph and application. For example, suppository baskets are manufactured from PTFE with vertical slits to facilitate dissolution.

Mesh size refers to the number of openings per linear inch and the diameter of the wire used to create the mesh is specified.

If the basket is reactive with the dosage form, then it can be coated with PTFE or an inert substance such as gold or titanium.

The diagram shows that the basket is connected to the rotating shaft by the use of clips onto a basket hub. The basket hub also has a specific dimension and importantly, a small vent hole at the top to allow air to escape from the basket should it become trapped. It is important that this hole does not become clogged over time.

Some baskets are supplied for use without clips and use a push-on O-ring instead. While these are excellent for many applications, it is important to show that the results from this design are the same as the results obtained with clips – a process known as equivalence. It is by no means guaranteed that the results will be the same in every case.

It is also important to note that different dissolution tester manufacturers will manufacture their baskets and hubs to slightly different specifications but all within the USP guidelines. Even though they look similar, baskets are not interchangeable between different tester manufacturers.

Note: Apparatus 1- Basket apparatus The assembly consists of the following:

- A covered vessel
- Transparent material
- A motor
- A metallic drive shaft,
- A cylindrical basket,
- The water bath permits the holding of temp inside vessel at 37°C ± 0.5°C
- The vessel is a cylindrical with hemispherical bottom.

Apparatus II (Paddle)

The paddle apparatus (Apparatus II) consists of a special, coated paddle that minimizes turbulence due to stirring. The paddle is attached vertically to a variable-speed motor that rotates at a controlled speed.

The tablet or capsule is placed into the round-bottom dissolution flask, which minimizes turbulence of the dissolution medium. The apparatus is housed in a constant-temperature water hall maintained at 37°C, similar to the rotating-basket method (Figure 4.2).

The position and alignment of the paddle are specified. The paddle method is very sensitive to tilting. Improper alignment may drastically affect the dissolution results with some drug products.

The same set of dissolution calibration standards is used to check the equipment before tests are run. The most common operating speed for Apparatus II is 50 rpm for solid oral dosage forms and 25 rpm for suspensions.

Apparatus II is generally preferred for tablets. A sinker, such as a few turns of platinum wire, may be used to prevent a capsule or tablet from floating.

A sinker may also be used for film-coated tablets that stick to the vessel walls or help position the tablet or capsule under the paddle. The sinker should not alter the dissolution characteristics of the dosage form.

The dimensions of the paddle are closely controlled. Any variations can easily have a detrimental effect on reproducibility from vessel to vessel. For this reason, paddles should be treated as a precision piece of equipment, stored correctly, and inspected regularly for scratches, dents or other imperfections.

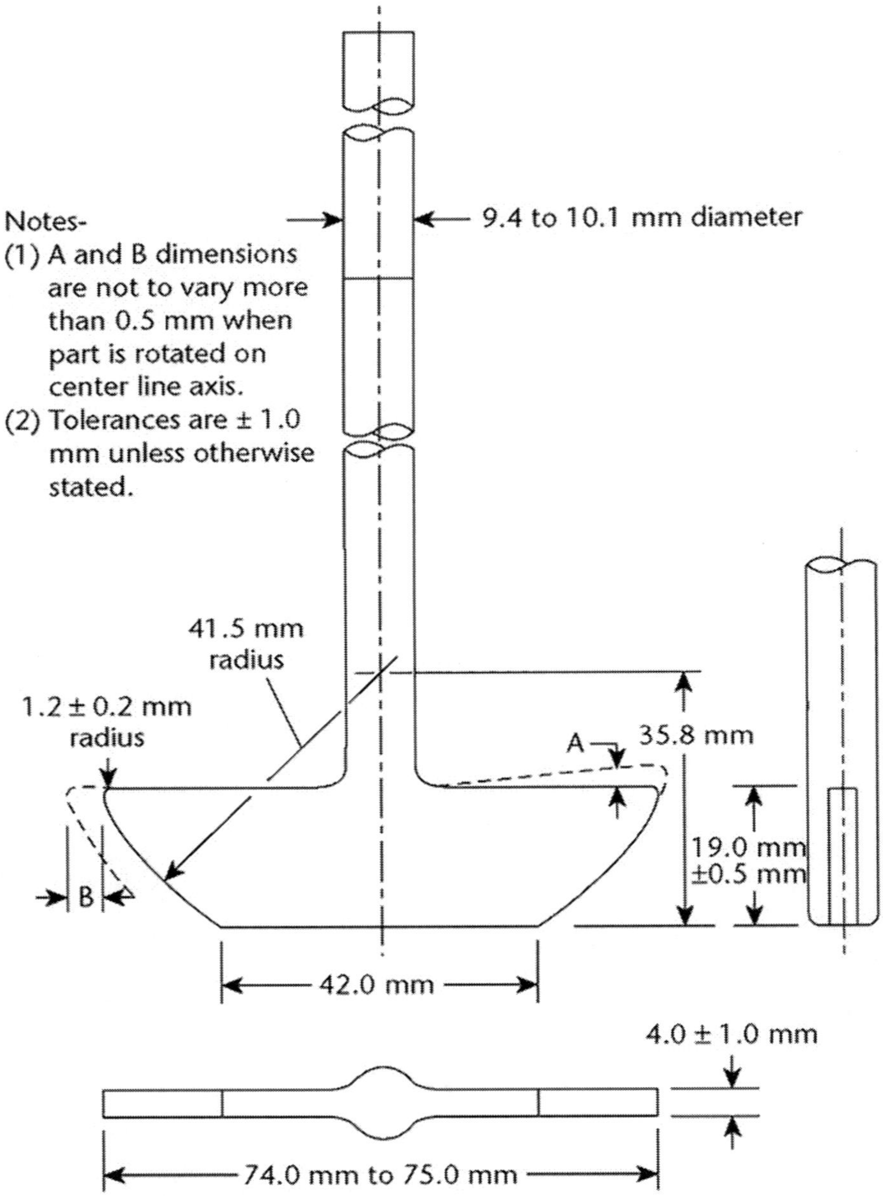

FIGURE 4.2 Dissolution apparatus II ({addle).

The dissolution test relies on uniform flow patterns and any imperfections can lead to turbulent flow. For that same reason, wobble and other deviations must be tightly controlled with specifications similar to the basket test.

Dissolution Medium

A solvent specified in the individual monograph or specific test procedure. Medium is deaerated by sparging with helium or warming the medium.

Procedure

Refer to the individual monograph or specific test procedure for dissolution parameters. Set up the dissolution apparatus as per the method.

Place the stated volume of dissolution medium into each of the dissolution vessels and equilibrate to 37°C$\pm$0.5°C, Ensure that the interior of the dissolution bath is clean, clear, and free of any obstructing factors (air bubbles, particles, etc.) that can interfere with the visual observation of product behavior during dissolution testing, Record the temperature range of the medium at the initial and final time points. Set the specified spindle speed. Set up the timer to monitor the dissolution time.

If using paddles, lower the paddles into the dissolution medium (paddles should not be rotating). Place one tablet or capsule into each of the six vessels. If capsules float, attach a small piece of non-reactive material, such as a few turns of wire helix or plastic capsule weights, to capsules. When a dosage unit is at the bottom of the vessel, start rotating the paddles and start the timer.

If using baskets, ensure that the baskets are dry and free of bubbles. Place 1 tablet or capsule into each of the six dry baskets, lower the baskets into the dissolution medium (baskets should not be rotating). When baskets are all the way down, start rotating the baskets and start the timer.

At the specified time intervals or at the end of the stated time, withdraw a sample from a zone midway between the surface of the dissolution medium and the top of the rotating basket or paddle, not less than 1 cm from the vessel wall using a cannula or a probe. Filter the sample solution through a filter specified in the method.

Where capsule shells interfere with the analysis, remove the contents of not less than six capsules as completely as possible and place the empty capsule shells in one vessel. Perform analysis as for the whole capsules and make necessary connections. If the dissolution apparatus consists of seven vessels, place all the empty capsule shells into the 7th vessel. Connection factors greater than 25% of the labelled claim are unacceptable.

Perform the analysis as directed in the individual monograph.

Acceptance Criteria for Individual Units

Each product has a specification which is known as Q.

There are three stages for Q:

S_1 stage:

At this stage, six individual capsules/tablets are tested. The individual results should be Q+5.

If all six results are equal or above Q+5 the test is passed at S_1 stage. If this criteria is not met, six more units are being tested.

S_2 stage:

Average of 12 units should be equal to or greater than Q, and no unit < Q – 15%. If this criterion is not met, 12 more units are being tested.

S_3 stage:

Average of 24 units should be equal to or greater than Q, and not more than 2 units < Q – 15% and no unit < Q – 15%.

Pooled Samples for Capsules. Uncoated Tablets and Plain Coated Tablets

Perform test for six individual units as described under Single Time Point Dissolution. Combine equal volumes of the filtered solutions of the six (or 12) individual solutions withdrawn and use the pooled sample as the test solution.

S_1 stage:

At this stage, six individual capsules/tablets are tested, and an equal amount of all six are combined and tested. The result should be Q + 10.

If this criteria is not met, six more units are being tested.

S_2 stage:

At this stage, six more individual capsules/tablets are tested and equal amount are combined. The average of the two sets should be Q + 5. If this criterion does not meet, 12 more units are being tested.

If all six results are equal or above Q + 5 the test is passed at S_1 stage. If this criteria is not met, six more units are being tested.

S_2 stage:

Average of 12 units should be equal to or greater than Q, and no unit < Q – 15%. If this criterion is not met, 12 more units are being tested.

S_3 stage:

Average of 24 units should be equal to or greater than Q.

Manual Dissolution Profile

Follow the procedure stated under the Single Time Point Dissolution, but withdraw a sample at every specified time interval (e.g., 5, 10, 15, 20, 25, and 30 minutes). Replace the withdrawn sample with an equivalent amount of dissolution medium after each sample is taken.

Testing of Samples

Samples are being tested using UV or HPLC.

In both instruments, a standard is prepared with the same concentration of the sample and analyzed.

An Example Method for Dissolution by UV

Dissolution by HPLC

Dissolution Medium: 0.01 N HCl (Dilute 8.5 mL Conc.HCl acid to 1000 mL pure water.

Apparatus:	Paddle
RPM:	100 RPM/min
Wavelength:	254 nm
Volume:	1000 mL
Temperature:	37.0°C ± 0.5°C
Run time:	30 minutes

Standard Stock Preparation

Accurately weigh 50 mg standard and transfer into a 100 mL volumetric flask. Dissolved and diluted to volume with Dissolution medium.

Standard Working Preparation

Pipette 4.0 mL standard stock to a 100 mL volumetric flask and dilute to volume with Dissolution medium.

Sample Preparation

Add 1000 mL dissolution medium to six vessels. Set up the heater to 37°C. Using a thermometer check the temperature of each vessel. When the temperature reaches 37.0°C ± 0.5°C, add one tablet to each vessel and start the rotation of the paddle. At 30 minutes, withdraw samples from each vessel and filter.

Procedure

Setup the UV instrument to 254 nm. Fill the UV cell with dissolution medium, place the cell in the designated place and zero the absorbance.

Read the standard working solution.
Read all six samples.
Read the standard working solution in the end.

Calculations

% Assay = Area of sample/Area of standard × weight of standard × dilution of sample/dilution of standard × 1/number of tablets × 100/label claim

Example Calculation by UV

A capsule containing 20 mg active (Potency of the standard is 99.9%) tested for dissolution, with 1000 mL dissolution medium. 50 mg of the standard was dissolved in 100 mL solvent, and 4.0 mL of the resulting solution was diluted to 100 mL. The final working standard and sample were analyzed at UV. The Q for this product is 80.

The following observations were recorded.

The absorbance of standard = 0.5001
The absorbance of Sample 1 = 0.4980
The absorbance of Sample 2 = 0.4875
The absorbance of Sample 3 = 0.4881
The absorbance of Sample 4 = 0.4581
The absorbance of Sample 5 = 0.4687
The absorbance of Sample 6 = 0.4974
The absorbance of Capsule blank = 0.0007

% Dissolved = Abs of sample/Abs of standard × weight of standard × dilution of sample/dilution of standard × 100/label claim

Sample dilution = 1000
Standard dilution = 100 × 100/4 = 2500
Corrected weight of standard for potency = 50 × 99.9/100 = 49.95 mg
Each absorbance of the sample will be subtracted from the capsule blank
% dissolved for sample 1:

= (0.4980 − 0.0007) × 0.5001 × 49.95 × 1000/2500 × 100 = 99.3%

The six results will be 99.3, 97.2, 97.4, 91.4, 93.5, 99.2%

The average of six = 96.3%. All individual results are more than Q (80) + 5, sample passed S_1 USP criteria.

Example Calculation by HPLC

A capsule containing 50 mg active tested for dissolution, with 900 mL dissolution medium. 55 mg (Potency of the standard is 99.8%) of the standard was dissolved in 100 mL solvent, and 10.0 mL of the resulting solution was diluted to 100 mL solvent. The final working standard and sample were analyzed by HPLC. The Q for this product is 75.

The following observations were recorded.

The Average area of standard = 2,299,221
The area of Sample 1 = 2,298,741
The area of Sample 2 = 2,187,452
The area of Sample 3 = 2,052,014
The area of Sample 4 = 1,600,181
The area of Sample 5 = 2,302,221
The area of Sample 6 = 2,014,421

% Dissolved = Area of sample/Area of standard × weight of standard × dilution of sample/dilution of standard × 100/label claim

Sample dilution = 900

Standard dilution = 100 × 100/10 = 1000

Corrected weight of standard for potency = 55 × 99.8/100 = 54.89 mg

% dissolved for sample 1:

$$= 2{,}298{,}741 \times 2{,}299{,}221 \times 54.89 \times 900/1000 \times 100 = 98.8\%$$

The six results will be = 98.8, 94.0, 88.2, 68.8, 98.9, 86.6%, Average = 89.2

The first criterion for S1 is that each unit must be Q + 5; since one of the results is 68.8%, the test does not meet S2 criteria, and six more tablets will be tested. If the average of 12 results is equal to or higher than Q, no unit < Q – 15%, the test will pass the S2 criteria.

QUESTIONS AND ANSWERS

1. **Define assay test.**
   Answer: The assay is an analytical procedure determining the presence of an active pharmaceutical ingredient in a dosage form.

2. **What is the difference between assay and content uniformity?**
   Answer: An assay test is done on a representative sample containing a composite sample of multiple dosages combined and tested.
   In Content uniformity, ten or more dosages are tested individually to determine the active content in a single dose.

3. **Average tablet weight of a 50 mg strength tablet is 200.1 mg/tablet. 800.2 mg powder of this tablet is dissolved in a 200 mL volumetric flask. 5.0 mL of resulted solution is diluted to 100 mL. 50.01 mg (The potency of the standard is 99.8%) standard is dissolved in a 100 mL volumetric flask. 10.0 mL of resulted solution is diluted to 100 mL. The average area of the standard was found to be 4,012,223, while the sample area was 3,991,114. Calculate the % Assay.**
   Answer: 99.3%

4. **Why is controlling impurities critical in a drug?**
   Answer: Because they can affect the product's safety and efficacy.

5. **Why in the impurity test sample prepared in high concentration?**
   Answer: Impurities are very low in concentration and challenging to quantify, so the sample is prepared in high concentration to increase the concentration of impurity in the solution.

6. **Average tablet weight of a 20 mg strength tablet is 150.2 mg/tablet. 600.6 mg powder of this tablet is dissolved in a 100 mL volumetric flask. 10.01 mg (The potency of the standard is 100.0%) standard is dissolved**

in a 200 mL volumetric flask. 4.0 mL of resulted solution is diluted to 100 mL. The average area of the standard was found to be 39,013, while the impurity area in the sample was 3912. Calculate the % impurity.
Answer: 0.025%

7. **Define the dissolution process.**
Answer: The release of a drug from a solid dosage in a certain medium and condition in the human body is called dissolution.
When a solid dosage like a tablet or capsule is taken orally, the medicine releases at a specific time and condition. This process is called dissolution.

8. **What are the two most common apparatuses of dissolution tests?**
Answer: Baskets and Paddles

9. **How many tablets or capsules are tested at S1 stage?**
Answer: 6 Tablets or capsules.

10. **A tablet containing 10 mg active tested for dissolution, with 900 mL dissolution medium. 23.0 mg (Potency of the standard is 99.0%) was dissolved in 100 mL solvent, and 5.0 mL of the resulting solution was diluted to 100 mL. The final working standard and sample were analyzed at UV. The Q for this product is 75. The following observations were recorded.**

The absorbance of standard = 0.4101
The absorbance of Sample 1 = 0.4080
The absorbance of Sample 2 = 0.4175
The absorbance of Sample 3 = 0.4081
The absorbance of Sample 4 = 0.4025
The absorbance of Sample 5 = 0.4145
The absorbance of Sample 6 = 0.4024
Is the result passing S1 Criteria?

Answer: Average = 102.1%, (V-1 = 101.9%, V-2 = 104.3%, V-3 = 102.0%, V-4 = 100.6%, V5 = 103.6%, V-6 = 100.5%). Each unit is more than Q+5, the average is more than Q. Results passes S1 Criteria.

5 Physical Testing of Raw Materials, Intermediate, and Finished Products

Physical tests are routine tests performed for raw material and in-process production. Depending upon the individual product monograph, some physical tests are also performed for the finished products. These tests are carried out before the manufacturing process to ensure that established product quality is met before they are approved for release.

The function of in-process quality control is monitoring the manufacturing processes to ensure that the product conforms to its specifications.

Physical tests during in-process are performed at regular intervals during a process step or at the end of a process step. The tests allow the production to identify and follow all changes that may occur during applied technological procedures. It gives the production security that the finished products fulfil all quality requirements.

In this chapter, some common tests will be discussed.

LOSS ON DRYING (LOD)

The procedure determines the amount of volatile materials of any kind defined as "Loss on Drying" value.

Unless otherwise directed in the individual monograph, conduct the determination of Loss on Drying on a 1- to 2-g test sample. Mix the substance and, if it is in the form of large particles, reduce the particle size to about 2 mm by grinding before weighing.

The weighing bottle is to be dried for about 30 minutes under the same conditions to be employed in the determination and cooled to room temperature in a desiccator. Place the test specimen in the bottle, replace the stopper, and accurately weigh the stoppered bottle and the contents.

Distribute the test specimen as evenly by gentle shaking. The depth of the sample should be about 5 mm generally, and NMT 10 mm in the case of low bulk density materials. Place the sample bottle in the drying chamber, remove the stopper, and leave it also in the chamber. Dry the test specimen at the specified temperature and time conditions as specified in the product monograph.

The Loss on Drying value is a function of both temperature and time. Therefore, these values must be identified and reported. The temperature specified in the monograph is to be regarded as being within the range of $\pm 2°$ of the stated value.

When "dry to constant weight" is specified in a monograph, drying should be continued until two consecutive weightings do not differ by more than 0.50 mg/g of

DOI: 10.1201/9781003366393-5

specimen taken, where the second weighing follows an additional hour of drying. Upon opening the chamber, reapply the same stopper to the bottle, and allow it to come to room temperature in a desiccator before weighing accurately.

Where capsules are to be tested, use a representative sample mixture, excluding the capsule shell, from not fewer than four capsules.

Where tablets are to be tested, use a representative sample mixture from NLT 4 crushed tablets.

Where drying under a vacuum over a desiccant is directed in the individual monograph, a vacuum desiccator or a vacuum drying pistol, or other suitable vacuum drying apparatus, has to be used.

Where drying in a desiccator is specified, exercise particular care to ensure that the desiccant is fully effective.

Where drying in a capillary-stoppered bottle under vacuum is directed in the individual monograph, use a bottle or tube fitted with a stopper having a 225 ± 25-μm diameter capillary, and maintain the heating chamber at a pressure of 5 mm or less of mercury. At the end of the heating period, admit dry air to the heating chamber, remove the bottle, and with the capillary stopper still in place allow it to cool to room temperature in a desiccator before weighing. Calculate the Loss on Drying (LOD).

Calculations

Note the temperature and record the drying time.

$$\% \text{ LOD} = \frac{W_1 - W_2}{W_1} \times 100$$

Where:

W_1 = Weight of sample before drying = (Weight of bottle, lid and sample before drying) – (Weight of bottle and lid before drying)

W_2 = Weight of the bottle after drying = (Weight of bottle, lid and sample after drying) – (Weight of bottle and lid before drying)

Karl Fischer Titration

For substances appearing to contain water as the only volatile constituent, the procedure is called "Karl Fischer method. The Karl Fischer (KF) titration is a method of determining the water content of solid, liquid, and gaseous samples. It is the technique preferred for use in industrial quality control. The principle of Karl Fischer titration is based on the oxidation reaction between iodine and sulphur dioxide. Water reacts with iodine and sulphur dioxide to form sulphur trioxide and hydrogen iodide. An endpoint is reached when all water is consumed.

The difference between Karl Fischer and the moisture analyzer is that the moisture content analysis provides the total amount of water but does not differentiate the type of water. Karl Fischer titrations are effective at quantifying even the tightly "bound" water and are often considered a better moisture analysis method than loss on drying.

DISINTEGRATION

The disintegration test determines whether tablets or capsules disintegrate within the prescribed time when placed in a liquid medium at experimental conditions presented in this method.

Complete disintegration is defined as that state in which any residue of the unit, except fragments of insoluble coating or capsule shell, remaining on the screen of the test apparatus or adhering to the lower surface of the disk, is a soft mass having no palpable firm core. Disintegration does not imply a complete solution for the unit or its active constituents.

Complete disintegration is defined as that state in which any residue of the unit, except fragments of insoluble coating or capsule shell, remaining on the screen of the test apparatus or adhering to the lower surface of the disk, if used, is a soft mass having no palpably firm core.

Apparatus

The apparatus consists of a basket-rack assembly, a 1000-mL low-form beaker 138–160 mm in height and having an inside diameter of 97–115 mm for the immersion fluid, a thermostatic arrangement for heating the fluid 35°C–39°C, and a device for increasing and lowering the basket in the immersion fluid at a constant frequency rate 29–32 cycles/min through a distance of NLT 53 mm and NMT 57 mm. The volume of the fluid in the vessel is such that at the highest point of the upward stroke, the wire mesh remains at least 15 mm below the surface of the fluid and descends to NLT 25 mm from the bottom of the vessel on the downward stroke. At no time should the top of the basket-rack assembly become submerged. The time required for the upward stroke is equal to the time required for the downward stroke, and the change in stroke direction is a smooth transition, rather than an abrupt reversal of motion. The basket-rack assembly moves vertically along its axis. There is no appreciable horizontal motion or movement of the axis from the vertical.

Basket-Rack Assembly

The basket-rack assembly consists of six open-ended transparent tubes, each 75.0–80.0 mm long and has an inside diameter of 20.7–23 mm and a wall 1.0–2.8 mm thick; the tubes are held in a vertical position by two plates, each 88–92 mm in diameter and 5–8.5 mm in thickness, with six holes, each 22–26 mm in diameter, equidistant from the center of the plate and equally spaced from one another. Attached to under the surface of the lower plate is a woven stainless steel wire cloth, which has a plain square weave with 1.8–2.2-mm apertures and with a wire diameter of 0.57–0.66 mm. The parts of the apparatus are assembled and rigidly held by means of three bolts passing through the two plates. A suitable means is provided to suspend the basket-rack assembly from the raising and lowering device using a point on its axis.

The design of the basket-rack assembly may be varied somewhat, provided that the specifications for the glass tubes and the screen mesh size are maintained. The basket-rack assembly conforms to the dimensions found in Figure 5.1.

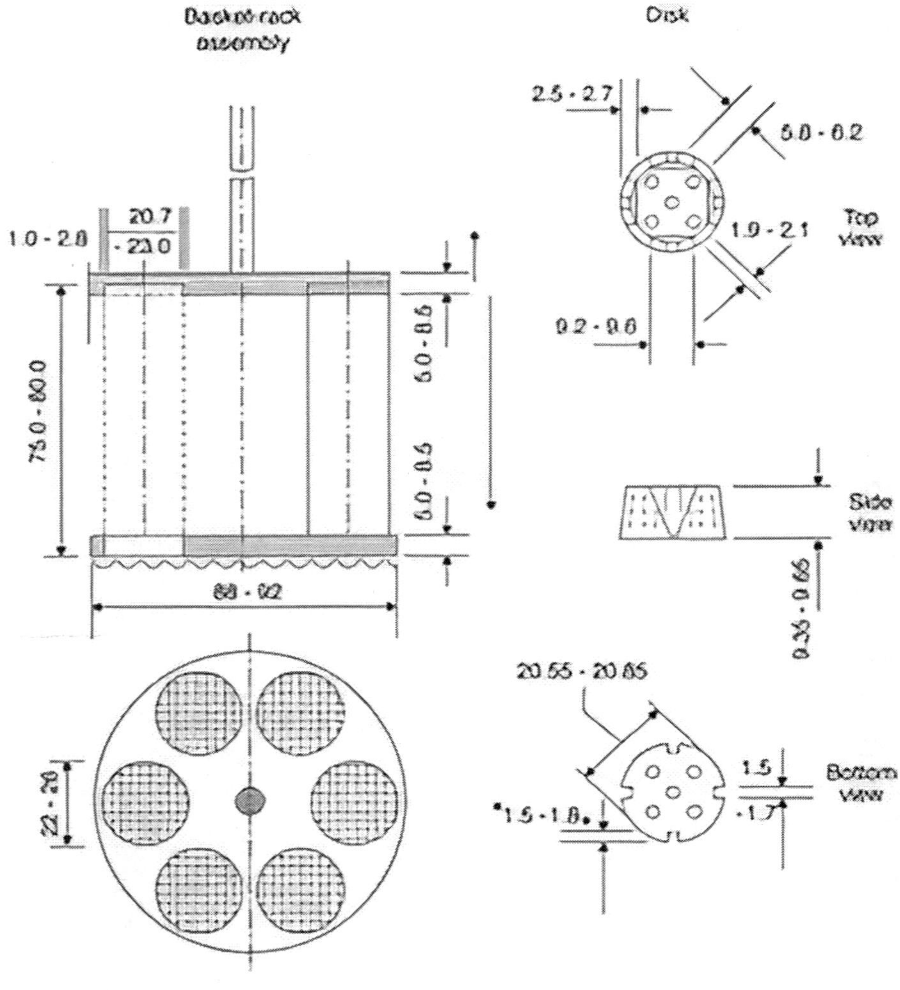

FIGURE 5.1 Disintegration apparatus, all measurements are in mm.

Apparatus

The use of disk is permitted only where specified in the specification or monograph. The use of automatic detection employing modified disks is allowed if the use of the disks is permitted.

USP Method

If the testing period is not prescribed, record the time when the first and the last tablets or capsules disintegrate completely.

Uncoated and Plain Coated Tablets

Place one tablet into each of the six tubes of the basket and add a disk if indicated in the monograph or specification.

Operate the apparatus using water or the specified medium as the immersion fluid, maintained at 37°C ± 2°C. At the end of the time specified in the monograph or the specification lift the basket from the fluid and examine the tablets. All the tablets should have disintegrated completely. If one or two tablets fail to disintegrate completely, repeat the test on 12 additional tablets. The requirement is met if not fewer than 16 of the total of 18 tablets are disintegrated completely.

Enteric-Coated Tablets

Place one tablet into each of the six tubes of the basket and if the tablet has a soluble external sugar coating, immerse the basket into the water at room temperature for 5 minutes. Then operate the apparatus using simulated gastric fluid TS maintained at 37°C ± 2°C, as the immersion fluid. After 1 hour of operation in simulated gastric fluid TS, lift the basket from the fluid, and examine the tablets. The tablets show no evidence of disintegration, cracking, or softening. Operate the apparatus with the disk, if specified, using simulated intestinal fluid TS maintained at 37°C ± 2°C, as the immersion fluid, for the time specified in the monograph or specification. At the end of the specified time, lift the basket and examine the tablets. All the tablets should have disintegrated completely.

If one or two tablets fail to disintegrate completely, repeat the test on 12 additional tablets. The requirement is met if not fewer than 16 of the total of 18 tablets are disintegrated completely.

Gelatin Capsules (Hard and Soft)

Follow the procedure as described above for Uncoated and Plain Coated Tablets, except, attaching a removable wire cloth, which has a plain square weave with 1.8–2.2 mm mesh apertures and with a wire diameter of 0.600–0.655 mm to the surface of the upper plate of the basket-rack assembly. Examine the capsules within the time limit specified in the individual monograph or specification. All of the capsules should have disintegrated except for fragments from the capsule shell. If one or two capsules fail to disintegrate completely, repeat the test on 12 additional capsules. The requirement is met if not fewer than 16 of the total of 18 capsules are disintegrated completely.

BP and Ph.Eur. Method

Uncoated Tablets

Introduce one tablet into each of the six tubes of the basket and add a disk to each tube. Suspend the assembly into the beaker containing water maintained at

37°C ± 2°C and operate the apparatus for 15 minutes, unless otherwise justified and authorized. Remove the assembly from the water and examine the state of the tablets. All the tablets should have disintegrated completely. If the tablet fails to comply because of adherence to the disk, repeat the test on additional six tablets omitting the disks. The requirement is met if all six tablets are disintegrated completely.

Coated Tablets

Introduce one tablet into each of the six tubes and add a disk to each tube. Suspend the assembly into the beaker containing water maintained at 37°C ± 2°C and operate the apparatus for 60 minutes (or 30 minutes for film-coated tablets), unless otherwise justified and authorized. Remove the assembly from the water and examine the state of the tablets.

If any of the coated or film-coated tablets have not disintegrated, repeat the test on additional six tablets, replacing water with 0.1M hydrochloric acid. The requirement is met if all six tablets are disintegrated in the acidic medium. If the tablet fails to comply because of adherence to the disks, repeat the test on additional six tablets omitting the disks. The requirement is met if all six tablets are disintegrated completely.

Dispersible Tablets

Follow the procedure for the disintegration of the uncoated tablets. Dispersible tablets disintegrate within 3 minutes (unless otherwise specified) using water maintained at 15°C–25°C as the liquid medium.

Enteric-Coated Tablets

Introduce one tablet into each tube, suspend the assembly into the beaker containing 0.1 M hydrochloric acid maintained at 37°C ± 2°C, and operate without disks for 120 minutes, unless otherwise stated in the individual monographs. Remove the assembly from the liquid. No tablets show signs of cracks that would allow the escape of the content or disintegration, apart from the fragments of coating. Replace the liquid in the beaker with mixed phosphate buffer solution pH 6.8, add a disk to each tube, and operate the apparatus for further 60 minutes. Remove the assembly from the liquid. All the tablets should have disintegrated completely. If the tablet fails to comply because of adherence to the disks, repeat the test on additional six tablets omitting the disks. The requirement is met if all six tablets are disintegrated completely.

Gelatin Capsules (Hard and Soft)

Introduce one capsule into each of the six tubes. Suspend the assembly into the beaker containing water maintained at 37°C ± 2°C or when justified and authorized, 0.1 M hydrochloric or artificial gastric juice may be used as the liquid medium. Operate the apparatus for 30 minutes, unless otherwise justified and authorized, remove the assembly from the liquid, and examine the state of the capsules. All the capsules should have disintegrated except for fragments from the capsule shell.

For Hard Capsules: If the capsules float on the surface, add the disk. For Soft Capsules: Add a disk to each tube. Liquid active substances dispensed in soft capsules may attack the disk. In these circumstances and where authorized, the disk may be omitted.

If the capsules fail to comply because of adherence to the disks, repeat the test on additional six capsules omitting the disks. The requirement is met if all six capsules are disintegrated completely.

pH Measurement

pH is a critical test for in-process and finished products.

Usually, it is tested for liquids and creams, but in some sold dosages, it is also performed.

pH is the inverse log of the hydrogen ion concentration in a liquid. The pH determines the acidity or basicity of a liquid.

The instrument used to measure the pH is known as pH meter. pH meter is commonly used in laboratories.

pH meter consists of a pH sensor, and pH signals can be transmitted digitally to an external device such as a computer.

pH Measurement System

The measurement system consists of three units:

1. Measuring Electrode
   It is sensitive to hydrogen-ion activity, typically a glass electrode, though other electrode types are possible.

2. Suitable Reference Electrode
   A reference electrode is used.

3. Voltage Measurement System
   A voltage measuring system with an input resistance capable of measuring at a high input impedance of the pH sensor.

The measuring and reference electrodes may be separated or combined. The voltage measurement system may be separated from the pH sensor or integrated into the sensor. For most applications, a temperature measurement will be necessary. A temperature device may be embedded into the pH sensor, or an external temperature device may be used (Figure 5.2).

Buffer Solutions for Calibration of the pH Measurement System

Buffer solutions for calibration can be prepared. Buffer salts of requisite purity can be obtained from the National Institute of Standards and Technology (NIST), other national authorities, or other suppliers. Buffer solutions should be stored in appropriate containers that ensure the stability of the pH through the expiry date and fitted with tight closure.

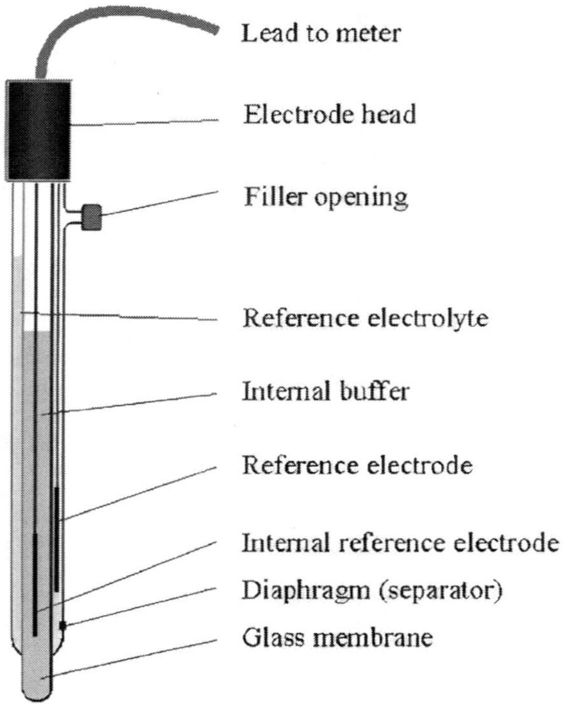

FIGURE 5.2 pH electrode.

Procedure

Electrode Preparation and Maintenance

Electrodes must be rinsed with distilled water and blotted, not rubbed dry, with lint-free tissues. Electrodes must not be exposed to air for more than 30 seconds.

Use internal filling solution Ag/AgCl

While not in use, electrodes are stored in the following storage solutions:

3.5M KCl/AgCl solution in water is to be used for storage of the electrode

Calibration

Because of variations in the nature and operation of the available pH measurement systems, it is not practical to provide universal directions for the calibration of the measurement system.

However, general principles can be followed.

Examine the electrodes, especially the reference electrode and electrolyte level, if a liquid electrolyte is used. If necessary, replenish the electrolyte supply, and observe other precautions indicated by the instrument and electrode manufacturers.

Commercially available buffer solutions for pH measurement systems, calibrated by methods traceable to the National Institute of Standards and Technology (NIST) or other national authorities, labelled with a pH value accurate to 0.02 pH units, may be used. Solutions prepared from ACS reagent grade materials or other suitable materials may be used, provided that the pH of the resultant solution is the same as that of the solution prepared from the NIST (or other national authorities) certified material. Buffer solutions greater than 12 should be used immediately or prepared using freshly boiled water and stored under conditions to minimize carbon dioxide absorption and ingress.

Rinse the electrode first with DI water and then with each specific buffer before placing it in the solution.

A two-point calibration can be performed. Three buffer solutions are to be used for calibration and verification. Two of the buffer solutions are used for the calibration process, with the pH of the testing material expected to fall within this range. The third buffer solution with intermediate pH is used to verify calibration accuracy. When the pH value needs to be used during the calibration process, always use the pH value stated in that buffer lot's Certificate of Analysis (COA), not the value on the bottle label.

The pH range between the two selected calibration buffer solutions should be kept narrow to ensure better accuracy of the pH reading. See below examples:

- Calibration using pH 1.00 buffer and pH 4.00 buffer, pH 1.68, or 2.0 buffer can be used for verification
- Calibration using pH 4.00 buffer and pH 7.00 buffer, pH 6.86 buffer can be used for verification
- Calibration using pH 7.00 buffer and pH 10.00 buffer, pH 9.18 buffer can be used for verification
- Calibration using pH 9.18 buffer and pH 12.00 buffer, pH

10.00 buffer can be used for verification

After completion of the two-point calibration process (before measuring the pH of the verification solution), verify that the pH slope and offset are within acceptable parameters.

Typical acceptable parameters are a slope of 90%–105% and an offset of $0 \pm 30\,\text{mV}$. Depending on the pH instrumentation, the pH slope and offset may be determined in software or by manual methods. If using manual methods, follow supplier instructions to calculate the pH sensor slope/offset. If these parameters are not within acceptable parameters, the sensor should be cleaned appropriately, replenished, serviced, or replaced, and the two-point calibration process shall be repeated.

Prepare the test material according to requirements in the monograph or according to specific procedures. If the pH of the test sample is sensitive to ambient carbon dioxide, then use purified water that has been recently boiled and subsequently stored in a container designed to minimize the ingress of carbon dioxide.

Rinse the pH sensor with water, and then with a few portions of the test material.

Immerse the pH sensor into the test material and record the pH value and temperature.

In all pH measurements, allow sufficient time for the stabilization of the temperature and pH measurement.

TABLET HARDNESS OR CRUSHING STRENGTH TEST

Tablet hardness testing is a laboratory technique used by the pharmaceutical industry to determine a tablet's breaking point and structural integrity.

Tablet hardness testing ensures that solid tablets are hard enough to sustain mechanical strains during various situations like storage and packaging, delivery, and handling by the user.

Hardness Tester

Electrically driven hardness testers are widely used to measure the crushing strength of tablets.

This equipment eliminates the operator variability encountered with manual hardness testers. Newer equipment with printers is also available.

A force of about 4 kg is the minimum requirement for a satisfactory tablet. The measurement is usually carried out using a minimum of ten tablets (Figure 5.3).

Solid tablets are possibly the most often used form of dosage for pharmaceuticals.

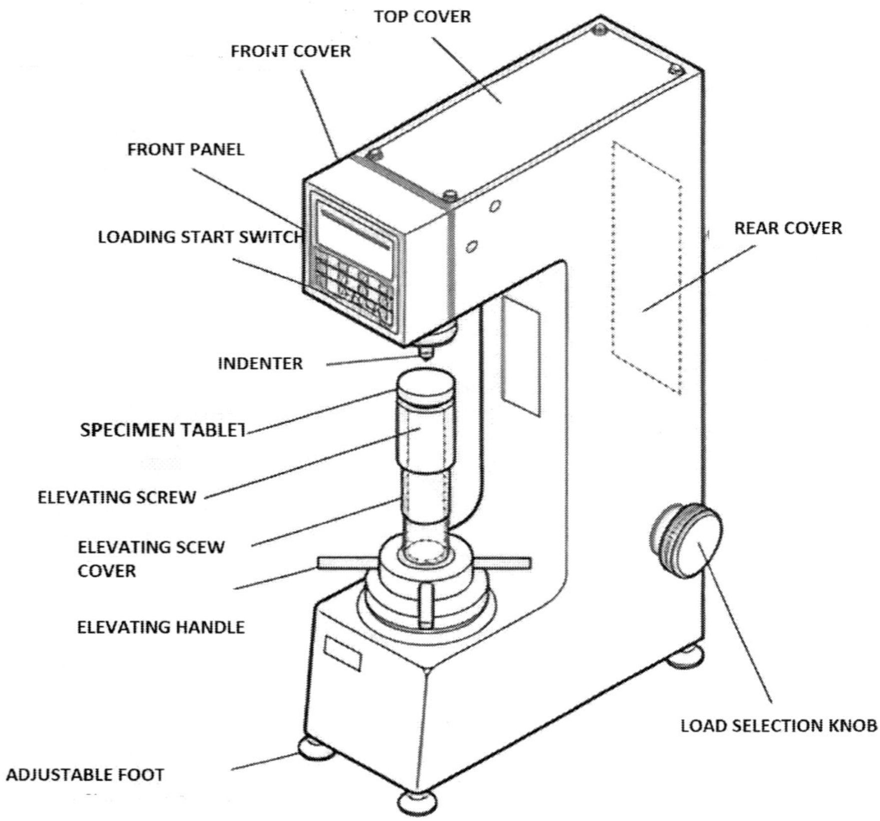

FIGURE 5.3 Hardness tester.

Tablet hardness acts as a basis to direct the development of products and as a quality-control requirement. Tablets should not be too soft or too hard.

Tablet hardness testing is a laboratory method which is applied to measure the breaking point of a tablet.

Most tablet hardness testers operate through a simple movement of a jaw or bar which squeezes the tablet between two rigid plates.

Thus, causing it to fracture under the pressure and hence, establishing the force needed.

There exist several varying tablet hardness testers, all operating around the same primary principle.

Hardness testers spun from simple handheld equipment to fully automated appliances.

Procedure

Place the tablet flat in contact with the stationary jaw as indicated below. Measure the hardness of ten individual tablets, unless otherwise specified.

Calculate the mean, range, and if required, the relative standard deviation of the ten tested tablets. The calculated mean value must fall within the acceptable range as outlined in the Product Specifications.

Placement of Tablets

Round Tablets: Place the tablet in contact with the stationary jaw, with score line (if present) perpendicular to the stationary jaw.

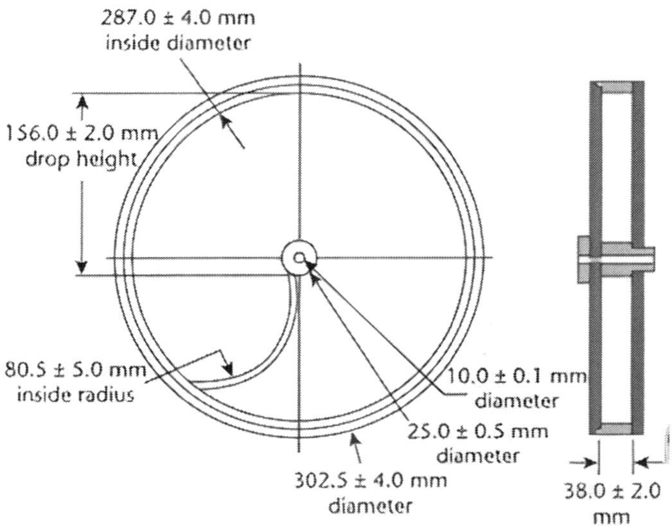

FIGURE 5.4 Friabilator.

Oval, Capsule Shaped, Elongated hexagon, and/or Rectangular Tablets: Place the tablet lengthwise, face down in contact with the stationary jaw.

Multi-edge (Polygonal) tablet: Place the tablets such that one flat edge is in contact with the stationary Tablets: jaw and a parallel flat edge is facing the moving jaw.

If the tablet shape does not correspond to any of the above shapes or conditions, refer to the specification of the product.

FRIABILITY

Tablet Friability is the physical condition that describes the tendency of tablets to break into smaller pieces or to detach a percentage of powder or powder loss from the tablet's outer surface under mechanical and physical stress.

During transportation and handling, tablets are subjected to mechanical shocks, which can lead to chipping and, in extremes, breaking, or capping. So, the tablet must be formulated to withstand this stress. It is also crucial that loosely packed powder does not erode from the tablets, which could cause a handling hazard.

A weak tablet could be due to several factors, including poor tablet design, low moisture content, insufficient binder, and over-lubrication.

The Friability tester is known as the friabilator. It consists of a drum, with an internal diameter between 283 and 291 mm and a depth between 36 and 40 mm, of a transparent synthetic polymer with polished internal surfaces and is subject to minimum static build-up. One side of the drum is removable. The tablets are tumbled at each turn of the drum by a curved projection with an inside radius between 75.5 and 85.5 mm that extends from the middle of the drum to the outer wall. The outer diameter of the central ring is between 24.5 and 25.5 mm. The drum is attached to the horizontal axis of a device that rotates at 25 ± 1 rpm. Drums with dual scooping projections or an apparatus with more than one drum for the running of multiple samples at one time are also permitted (Figure 5.4).

If a single tablet weight is less than 650 mg, take 6.5 g tablets or as near as possible to 6.5 g. For tablets with more than 650 mg/Tablet, take 10 tablets and weigh them.

The tablets should be carefully deducted before testing with air pressure or a soft brush. Accurately weigh the tablets and place them in a friabilator wheel. Rotate the wheel 100 times (i.e. approximately 4 minutes), unless otherwise specified and remove the tablets. Remove any loose dust from the tablets as before, and accurately weigh. In the case of hygroscopic tablets, a humidity-controlled environment is required for testing. Calculate the percentage of Loss in Weight.

If tablet size or shape causes irregular tumbling, adjust the drum so that the base forms about a 10° angle with the bench top and the tablets no longer bind together when lying next to each other, which prevents them from falling freely.

$$\% \text{ Loss in weight} = \frac{W_1 - W_2}{W_1} 100$$

W_1 = Net Weight (g) of tablets before
W_2 = Net Weight (g) of tablets after

If cracked, cleaved, or broken tablets are present in the tablet sample after tumbling, the sample fails.

QUESTIONS AND ANSWERS

1. **A 10.3256 g glass container was used to determine LOD. Container was dried for 30 minutes at 100°C. Sample was crushed in small particles and placed in the same container. The weight of sample and container was found to be 11.6521 g. The container containing the sample was dried for 4 hours at 100°C. The weight after drying was 11.5896 g. The specification for the test is "Not more than 5.0% for 4 hours at 100°C. Calculate the %LOD and determine if the test meets the specification.**
   Answer:
   Weight of empty container = 10.3256 g
   Weight of empty container + sample before drying = 11.6521 g
   Weight of sample before drying = 1.3265 g
   Weight of empty container + sample after drying = 11.5896 g
   Weight of sample after drying = 1.2640 g

$$\% \text{ LOD} = \frac{(1.3265 - 1.2640)}{(1.3265)} \times 100 = 4.7\%$$

   The test meets the specification and passes.

2. **How long, and in what condition is the weighing apparatus to be dried before weighing the original sample?**
   Answer: The weighing bottle is to be dried for about 30 minutes under the same conditions to be employed in the determination.

3. **How much sample should be weighed to determine the LOD?**
   Answer: 1–2 g of sample powder or as specified in the product monograph.

4. **For the disintegration test what should be the temperature of the water bath?**
   Answer: 15°C–25°C for dispersible tablets and for others 37°C ± 2°C.

5. **How many tablets/capsules are used for initial testing for disintegration test.**
   Answer: Six tablets or capsules.

6. **What is considered as "Complete disintegration"?**
   Answer: Complete disintegration is defined as that state in which any residue of the unit, except fragments of insoluble coating or capsule shell, remains on the screen of the test apparatus or adhering to the lower surface of the disk

7. **What are major parts of pH meter?**
   Answer:
   1. Measuring Electrode
   2. Suitable Reference Electrode
   3. Voltage Measurement System

8. **If the pH of a solution is to be adjusted to 7.2 which two buffer will be used to calibrate the pH meter?**
   Answer: Buffer 4.00 and 7.00

9 **How will you define the hardness?**
   Answer: Determine the breaking point and structural integrity of a tablet.

10. **How and oval shaped tablet is placed at hardness tester?**
    Answer: It should be placed the lengthwise, face down in contact with the stationary jaw

11. **For friability test, if the weight of an individual tablet is less than 650 mg how much sample to be weighed?**
    Answer: 6.5 g

12. **A friability test was performed, the initial weight of the tablets was 6.6329 g. After rotating for 100 times, the weight remained at 6.6013 g, and a broken tablet was found. The limit of the test is not more than 0.8%. Calculate the % loss. Is the test passed?**
    Answer:

$$\% \text{ loss} = \frac{(6.6329 - 6.6013)}{6.6329} \times 100$$

$$= 0.48\%$$

The result is well within the limit of friability, since a broken table is found overall results failed.

6 Data Integrity

Regulatory Bodies all around the world have been observing a critical issue, Data Integrity. Massive problems were found during inspections by the pharmaceutical regulatory agency because mal practices had given rise to the product for patients that was not safe for them to use. Data is established when various information and results are collected together; however, this data is not much useful without integrity. Accuracy and original data increase the chances of stability and performance of an organization.

In the simplest words, Data Integrity could be explained as follows;

The raw data has to be complete, reliable, and accurate and no changes or deletions could be introduced without appropriate authorization.

Authentication of Data Integrity is a critical part of the FDA's mission to ensure the safety, efficacy, and quality of human and veterinary drugs, biological products, and medical devices. As such, the FDA expects that all data which is submitted to the agency must fulfill the definition of completeness, consistency, and accuracy and should be attributable, legible, contemporaneously recorded, original or a true copy, and reliable.

Data integrity must not be mistaken as data security. The difference between the two terminologies should be clearly understood. Data Security refers to the prevention of data corruption through the use of controlled access mechanisms as well as ensuring that data is accessed only by its assigned users, thus ensuring the privacy and protection of data.

Whereas Data Integrity refers to the quality of data, which assures that the data is complete and has a whole structure and its prime motive is the validity of data.

SIGNIFICANCE OF DATA INTEGRITY IN GMP

Data Integrity plays a key role in all areas of GMP compliance and Health regulators expect data to be meaningful, taking into consideration the design, operation, and monitoring of systems and controls based on risk to patient, process, and product.

Data Integrity is highly important in any industry, but in pharmaceutical industry, its significance is further enhanced as any error in data could lead to serious consequences.

Principles of Data Integrity had to be standardized in order to regulate them and achieve better processes and higher quality products. This leads to ALCOA principle.

ALCOA is an acronym for the original five principles of Data Integrity that should be possessed by data in order to maintain the integrity.

The acronym "ALCOA" defines that data should be:

1. Attributable
2. Legible

DOI: 10.1201/9781003366393-6

3. Contemporaneous
4. Original and
5. Accurate

ALCOA relates to all sorts of data, whether paper or electronic, and these principles should be part of our data lifecycle, GDP, and Data Integrity initiatives.

It helps in developing strategies so that the integrity of the evidence is maintained both in research and manufacturing. The aspects of ALCOA in pharmaceuticals have been discussed below:

Attributable: To ensure accountability, the data should be attributable. This means that every piece of data entered into the record must be capable of being traced back to the person who has collected it. In other words, data should clearly demonstrate who observed and recorded it, when it was observed and recorded. This would further clarify that the person who recorded or entered the data was qualified, experienced, and trained accordingly to perform the desired responsibilities. Any changes made to records, for example corrections, deletions, changes, etc., will also be governed by this very principle.

Legible: This principle particularly applies on paper records that they should be legible. The data should be readable, easy to understand, and recorded permanently, and allows a clear picture of the sequencing of steps and original entries should be preserved and non-erasable and are reliable throughout the data lifecycle. This is very critical in the pharmaceutical industry as a mistaken spelling could result in the administering of a completely different drug.

For an electronic record to be considered legible, traceable, and permanent, creation of data in temporary memory must be avoided as well as immediately committing data to a permanent memory before moving on.

Contemporaneous: Contemporaneous could be explained as follows:

"Data should be recorded as it was observed and exactly at the time it was executed".

Basically, this factor provides the indication that the actions, events or decisions taken were recorded as they took place. This documentation can provide an accurate verification of what was done, or what was decided and what influenced the decision at that time.

Original: Source data should be accessible and preserved in its original form.

No matter whether the data paper based or electronic, the source in either case is referred to as source data or primary data. For example, if the data is related to validation, it should be originally recorded in an approved protocol or form or in a dedicated notebook and hence it should must remain available in that state. The basic purpose is to preserve the content and meaning of the original data where it is generated. Recording results in a notebook for transcription later can introduce errors and if your original data is handwritten and needs to be stored electronically, ensure that a "true copy" is generated, and the copy is verified for completeness and then migrated into the electronic system.

Accurate: Data accuracy means ensuring the recorded data is free from errors, complete, truthful, and reliable and conform to the protocol. If the data is edited without logging, then it means that the accuracy of data is lost. Therefore, editing should not be performed without documenting and annotating the amendments. It is of vital importance to always record who, when, and why changed the data record. ALCOA in pharmaceuticals helps both the companies and the users making it sure that there are no record-keeping errors due to which any product which is compromised on quality, efficacy, and safety is released onto the market.

Many of the concepts for ALCOA have been captured in the regulations as far back as 1978. Since that time, the industry has changed dramatically. The generic-drug industry has grown and in the United States alone accounts for more than 80% of the prescriptions written today. Coupled with the emergence of biosimilars, virtual companies, contract manufacturing organizations, rapid advancements in automation and information technology, and the globalization of the industry have resulted in reinterpretation of the attributes associated with maintaining the integrity of data throughout the product lifecycle, whether those data are generated from electronic, paper-based, or hybrid systems.

ALCOA+

The acronym ALCOA+ adds the concepts that, in addition to ALCOA, data also needs to be complete, consistent, enduring, and available. It is important to understand what each element of ALCOA+ means in order to apply the concepts appropriately with respect to a company's records. So when we talk of ALCOA+, we add up more values to original set ALCOA as follows;

a. Complete: Information that is critical to recreating and understanding an event. This would include any repeat or reanalysis performed on a laboratory test sample.
b. Consistent: The data is presented, recorded, dated, or time-stamped in the expected and defined sequence.
c. Enduring: The data or information must be maintained, intact, and accessible throughout their defined retention period.
d. Available: The data or information must be able to be accessed at any time during the defined retention period.

In order to minimize Data Integrity risks, the following actions are recommended:

1. Follow a software development lifecycle
2. Ensure all computer systems are compliant to 21 CFR Part 11 requirements

3. Implement error detection software
4. All computerized systems must be completely validated and qualified
5. Audit trails should be implemented effectively
6. There should be a backup and recovery procedure in place
7. The records should be secured with limited system access
8. The physical and logical security of systems must be protected
9. Design a Quality Management System with SOPs and logical controls
10. Establish a vendor management qualification program
11. Properly train users and maintain training records
12. Conduct internal audits to evaluate controls and procedures

QUESTIONS AND ANSWERS

1. **What is Data Integrity?**
   Answer: Data Integrity could be explained in the following words;
   "the raw data has to be complete, reliable, and accurate and no changes or deletions could be introduced without appropriate authorization".

2. **Define and explain ALCOA principles**
   Answer: ALCOA is an acronym for the original five principles of Data Integrity that should be possessed by all sorts of data, whether paper or electronic, in order to maintain the integrity.
   The acronym "ALCOA" defines that data should be;
   Attributable: This means that every piece of data entered into the record must be capable of being traced back to the person who observed and recorded it and when it was observed and recorded. Also, changes made to records, for example corrections, deletions, changes, etc., are also attributed to the responsible person.
   Legible: Legible means that the paper records should be readable, easy to understand, and recorded permanently, and allows a clear picture of the sequencing of steps and original entries should be preserved and not erasable and are reliable throughout the data lifecycle.
   The electronic records must always be created in a permanent memory to be considered as legible.
   Contemporaneous: This factor provides the indication that the actions, events, or decisions taken were recorded as they took place. In other words data should be recorded as it was observed and exactly at the time it was executed.
   Original: Original means the source data whether paper based or electronic should be accessible and preserved in its original form. For example, the data should be always recorded in an approved protocol or form or in a dedicated notebook and hence it should must remain available in that state so that the content and meaning of the original data where it is generated, could be preserved.
   Accurate: This principle means that the recorded data is free from errors, complete, truthful and reliable, and conform to the protocol. The data

should not be edited without logging in and documenting and annotating the amendments. It is of vital importance to always record who, when, and why changed the data record.

3. **What is an Audit trail?**
   Answer: In pharmaceutical industry, audit trail is a regulatory requirement. It is a series of records of computer events, about an operating system, an application, or user activities. Audit trail also contains in the record the name of persons who performed any activity with the date and time. Audit trail assures the record integrity and security.
   A computer system may have several audit trails, each devoted to a particular type of activity.

4. **What is the difference between ALCOA and ALCOA+?**
   Answer: The term ALCOA is an acronym, which stands for Attributable, Legible, Contemporaneous, Original, and Accurate. Alcoa was then expanded to Alcoa Plus (Alcoa +), by the addition of a few more concepts which are Complete, Consistent, Enduring, and Available.

5. **How does Data Security differ from Data Integrity?**
   Answer: There is a difference between the two terminologies that could be explained as follows;
   Data Security refers to the prevention of data corruption through the use of controlled access mechanisms as well as ensuring that data is accessed only by its assigned users, thus ensuring the privacy and protection of data.
   Whereas data integrity refers to the quality of data, which assures the data is complete and has a whole structure and it's prime motive is the validity of data.

7 Process Validation

According to the ICH guidelines Q7, Process validation can be defined as documented evidence that the process, operated within established parameters, can perform effectively and reproducibly to produce a medicinal product meeting its predetermined specifications and quality attributes.

Before stepping into the details of Process validation, let us understand first why Process Validation is important and why was it introduced to Pharmaceutical Good Manufacturing practices.

Process Validation concept does not have a very old history as other good manufacturing practice (GMP) elements. It was only in **1987** when US Food and Drug Administration released the process validation guidelines for the first time and introduced the pharmaceutical world with the term "Validation".

Although initially the pharmaceutical products were manufactured under some kind of process controls yet the controls were not strict enough to ensure and control the product quality to 100% and the manufacturing processes were still susceptible to produce medicines with compromised quality which were given to patients directly, resulting in obvious adverse consequences. These consequences led to the introduction of Quality Control department and various analytical functions related to it.

Establishing the regulatory agencies such as the US FDA, clinical testing prior to approval of new products, GMPs, and other controls were also taken into measures to prevent safety and efficacy adversities.

With the ever-growing pharmaceutical industry and complexity of the processes, it became apparent that a reactive approach of testing the products was not adequate. A proactive approach was then required to be applied and greater emphasis needed to be placed to control the processes prior to the finished stage of the products.

It was realized in the very beginning of biological products that many important quality attributes of the product were not directly measurable for release testing and that these attributes were critical to the product's safety and efficacy. It was therefore understood that a biopharmaceutical product is significantly defined by its manufacturing process and thus the process needed to be both tightly controlled and validated.

This was the beginning of the concept of Process Validation.

The fundamental approach was testing the process to assure it worked, along with periodic retesting of the manufacturing process to assure it was continuing to work. The US FDA thus came out with the definition of Validation as follows:

> PROCESS VALIDATION is establishing documented evidence which provides a high degree of assurance that a specific process consistently produces a product meeting its predetermined specifications and quality attributes

DOI: 10.1201/9781003366393-7

VALIDATION TEAM

Process Validation is one of many significant activities that is performed under the umbrella of Quality Assurance. Basically, validation is the fundamental part of Quality Assurance that induces the quality in the product by controlling each and every manufacturing step. It is a step-by-step procedure that is designed to ensure that a manufacturing process can consistently produce quality products. The pharmaceutical company's Quality Assurance department must establish a validation team comprising of a member each from Product development, Production, engineering and Quality Assurance itself, led by a QA personnel, preferably the Validation Manager. Process validation starts as early as the pilot batches manufacturing. This means it has to be done before releasing a new product, when applying any change on an existing product, and for periodically verifying the process. Established at the onset, a protocol should specify how the validation process will be carried out, including the parameters to be monitored, the samples to be taken, and the results to be accepted.

Process Validation comprises of the following steps:

1. Process Design: During this stage, the manufacturing process is defined in a way that it can reproduce the delivery of a medicinal product that meets predetermined specifications and quality attributes. To achieve this, the validation team should have a clear grasp of how the process actually works. In short, the process design can be further categorized into the following steps.

   - Risk assessment is carried out to identify the critical process and quality attributes of the product.
   - Protocols are written, reviewed, and approved.
   - Pilot batches and when appropriate, scale-up batches are manufactured.
   - Process is validated under the approved protocol using the sampling and testing criteria which were derived during the risk assessment step.
   - Report is written where critical quality attributes of the product and critical process parameters are defined to be monitored in the next phase which is process qualification.

2. Process Qualification: In the process Qualification stage, process design evaluation is performed to determine the effectiveness for quality production.

   The following steps will be taken in the process Qualification stage:

   - Ensure that the manufacturing facility was designed as per the requirements of current good manufacturing practice (CGMP).
   - Secondly, all utilities and equipment associated with the manufacturing should be qualified and validated to ensure that they are built and installed in compliance with design specifications.
   - Finally, process performance qualification should be executed through a protocol and documented in a report.

   Validation Protocol: A validation protocol must cover the following aspects;

a. Introduction of the product and process, Objectives of the validation, and Scope of the validation
b. Define the responsibilities of Production, Quality Assurance (QA), and Quality Control (QC) and Prerequisites (e.g., manufacturing records)
c. Process Validation Study Plan based on Quality Risk Management (QRM)
d. Establish a Validation Team that may be comprised of a person each from Research and Development, Engineering, Production, QA, and QC
e. List of Materials and Master Formula
f. List of Equipment
g. Process Flow Chart
h. Describe briefly the Manufacturing Process and identification of critical process parameters
i. In-process Test and Specifications or critical quality attributes
j. Sampling, Testing Plan, sample sizes, and Acceptance Criteria
k. Yield Details for at least three consecutive batches
l. Analytical Method Validation
m. Packaging process
n. Finished Product Specification
o. Review of Out-of-Specification Test Results, Deviation, and Change Control
p. Revalidation Criteria (if applicable)
q. Summary and Conclusion

3. Process Verification: Process verification is a continued process that comes after process design and process qualification. This stage deals with continuous monitoring of the validated process to ensure that the process remains in a state of control and within the validated parameters.

   Continued process verification often incorporates the use of statistical process control, the continuous monitoring and sampling of process parameters and quality attributes, and the scheduled maintenance of the facility, utilities, and equipment.

Types of Process Validation:
There are four types of process validation:

1. Prospective validation
2. Retrospective validation
3. Concurrent validation
4. Revalidation

Prospective Validation: Prospective validation is implemented when a new product is planned to be manufactured for the first time. This could be with a new formula or within a new facility. Prospective validation is always carried out before beginning commercial production batches. Prospective validation is actually the base of any new product that will be released under new conditions.

Retrospective Validation: This type of validation study is conducted only when at the time of launching a new product; a documented validation has not been carried out. This study is then fulfilled with the use of historical data and trend analysis to provide evidence that the process, despite not formally validated, is still at a state that it is intended to be in. This approach, however is no longer acceptable to process validation because it's a regulatory requirement that all new products should have already been validated before their commercial distribution.

Concurrent Validation: Concurrent validation is done under special circumstances during regular pharmaceutical production. Though, the approach is not widely practiced yet still acceptable under certain conditions. For example, manufacturing medically necessary drugs in coordination with the USFDA to prevent a short supply or a smaller batch size required immediately in the market and a prospective validation with three batches may result in scrapping the product unused, etc. Health regulators nevertheless continue to emphasize that it should only be used rarely.

Revalidation: Revalidation is more widely used for medical devices than drug products. It is executed when prospective validation reaches a conclusion that the manufacturing process is unable to produce the product consistently. Moreover, a criterion for revalidation may be indicated in the original validation protocol. The revalidation process may not be as comprehensive as the initial validation, especially if the situation only calls for some aspects to be repeated.

QUESTIONS AND ANSWERS

1. **Define process validation**

   Answer: PROCESS VALIDATION is establishing documented evidence which provides a high degree of assurance that a specific process consistently produces a product meeting its predetermined specifications and quality.

2. **Define and explain Process Design**

   Answer: Process design is the first stage of process validation. In this stage, the manufacturing process is defined in a way that it can reproduce the delivery of a medicinal product that meets predetermined specifications and quality attributes.

3. **How many steps are there in the process Design**

   Answer: The process design is further categorized into the following five steps.

   - Risk assessment is carried out to identify the critical process and quality attributes of the product.
   - Protocols are written, reviewed, and approved.
   - Pilot batches and when appropriate, scale-up batches are manufactured.

- Process is validated under the approved protocol using the sampling and testing criteria which were derived during the risk assessment step.
- Report is written where critical quality attributes of the product and critical process parameters are defined to be monitored in the next phase which is process qualification.

4. **What is process Qualification stage?**
Answer: Process Qualification is a stage where process design evaluation is performed to determine the effectiveness for quality production.

5. **How many steps are there in the process qualification stage?**
Answer: The following steps will be taken in the process Qualification stage:

- Ensure that the manufacturing facility was designed as per the requirements of current good manufacturing practice (CGMP).
- Secondly, all utilities and equipment associated with the manufacturing should be qualified and validated to ensure sure that they are built and installed in compliance with design specifications.
- Finally, process performance qualification should be executed through a protocol and documented in a report.

6. **What step follows the Process Qualification?**
Answer: Process verification follows process qualification which is a continued process.

7. **What is the significance of Process Verification?**
Answer: Process verification is a continued process which deals with continuous monitoring of the validated process to ensure that the process remains in a state of control and within the validated parameters.

8. **How does Prospective validation differ from Concurrent validation?**
Answer: Prospective validation is implemented when a new product is planned to be manufactured for the first time with a new formula or within a new facility. This type of validation is carried out before beginning commercial production batches. Whereas, concurrent validation is allowed only under special circumstances such as manufacturing medically necessary drugs in coordination with the USFDA to prevent a short supply or when a smaller batch size is required immediately in the market. However, a concurrent validation strategy is not encouraged by health regulators and should only be used rarely.

7.1 Protocol Generation

The process validation protocol is actually a written plan that describes the process to be validated. It further includes the list of production equipment and the entire approach as to how validation will be conducted. A validation protocol is necessary to define the specific items and activities that will constitute a validation study.

The validation protocol must be written by a qualified person of the concerned department and approved by appropriate signatories before the validation activities start.

A typical validation protocol should be comprised of the following segments;

Purpose/Objective: The aim of the validation study needs to be described under Purpose/Objective.

For example, the objective of this validation protocol is to provide documented evidence to demonstrate that the manufacturing process for (Name of Product), when carried out within defined parameters, will consistently perform its intended function and meet all pre-determined specifications and quality attributes.

Scope: The area or department where the protocol is concerned as well as the product's name will be mentioned.

Introduction: In this section a brief description of product is provided. Description will include the formulation of the product. The formulation can be presented in a list with the raw material code, description, and total quantity in the batch as follows:

Raw Material Code	Description	Quantity Per XXX L Batch	Label Claim
Comments:			

Besides the formulation of the product, details of the raw materials and manufacturing equipment are also enlisted in the introduction section of the protocol.

Raw Materials:

Raw Material	Code	Lot #	Supplier

DOI: 10.1201/9781003366393-8

Manufacturing Equipment:

Equipment	Serial Number

The number of rows in the tables above are only for example. The actual number of raw materials and equipment will usually be more and can be added accordingly.

PROCESS

Process Overview

Under Process Review, a brief summary of manufacturing steps in the manufacture of the product is given with the help of approved compounding instructions. The following important information is provided;

- The delivery of material from dispensing to different equipment as per the process requirements is given.
- For further information regarding the manufacturing process for the product, refer to the master document.

Validation Strategy

Three consecutive, successful production trials for each representative process/product will be necessary in order to consider the process/product validated.

The batches of XYZ 1 mg Tablets will be manufactured in ABC manufacturing area. These batches will be used for the above-mentioned type of process validation study. This work focused on the evaluation of the following steps:

- Review of the manufacturing formula and method, packaging instructions, raw material and finished product specifications and related SOPs.
- Issuance of Raw Material and Equipment used during manufacturing.
- Mixing time and mixer speed at different manufacturing steps.
- Temperature of vessel during mixing.
- The routine in-process control data and the critical parameters according to the manufacturing formula and method (MFM)
- Manufacturing and final batch yield.
- Critical packaging parameters.
- Physicals, Chemical, and microbiological properties of the active ingredients for the finished product

PROCEDURE AND PLANNED ACTIVITIES

- Define the procedures to be used in the validation
- Define planned activities in detail for the validation

Planned Activity	Responsibility
Manufacturing of Batches	Manufacturing Department
Sampling from Blender/Tank/Bin	Validation Specialist
Additional Sampling from line/Press	Validation Specialist/QA
Testing of all samples	Quality Control Department/Microbiology Department
Issuing final Validation report	Validation Specialist
Circulation of final report for approvals	Validation Specialist

DOCUMENTATION

Check the following in use documents before the start of process validation.

Documents Name and Title	Reference No.	Issue Date
• Manufacturing Formula and Method (MFM)		
• Master Packaging Instructions		
Finished Product Specification		
In-process Specification (Manufacturing)		
In-process Specification (Packaging)		

Raw Material Specifications

Provide the raw material specifications below in a similar fashion like above.

CRITICAL PROCESS PARAMETERS (MANUFACTURING)

Formulation (Active/Inactive Ingredients)

Composition of Product

Scale Weight in mg/tablet	Item No.	Material's Name	Code	Quantity/Batch XXX kg (100,000 tablet)	QC Status
	1	Active Pharma Ingredient			
	4	Lactose Monohydrate			
	5	Microcrystalline Cellulose			

(Continued)

Scale Weight in mg/tablet	Item No.	Material's Name	Code	Quantity/Batch XXX kg (100,000 tablet)	QC Status
	6	Sodium Starch Glycolate (Primojel)			
	7	Crospvidone (Kollidon CL)			
	8	Polysorbate 80 (Tween 80)			
	9	Iron Oxide Red			
	10	Magnesium Stereate			
	11	Purified water			

Names of the raw materials in the list above are added as an example of any given tablet based on the fact that most of the solid dosage forms have generally the same excipients. However, the table is only to give an idea as to how to create the list of materials for the protocol.

Issuance of Raw Materials

Re-check the quantities of all active and in-active raw materials for the following three consecutive batches before the start of manufacturing.

Batch No.	Issued On	Observation

7.2 Equipment Used

Check the following details of major equipment which are to be used during manufacturing.

Sr No.	Equipment Name	Manufacturer's Name	IQ/OQ Completion Date	Model	Serial No
1					
2					
3					
4					

Process Flow	Control Variables	Measured Responses
Addition of Raw Materials Active and Excipients	* Weighing / Sampling * Warehouse / QA	* Released by QC
Sizing Sifter Milling / Sieving	* Screen Type * Screen Size	*Granules Size Distribution
Mixing Dry and Wet mixing Addition of granulating fluid	* Mixing Time * Mixer Speed	
Drying Oven / Fluid Bed	*Drying Time *Drying Temperature	*LOD *Moisture Content
Blending Double Cone Blender / Tumbler Addition of Raw Materials Binders / Lubricants / Disintegrants	*Load Size *rpm *Blending Time	*Blend Uniformity *Flow Characteristics
Compressing High Speed Rotary Machine	*Compression Rate *Granules Feed Rate *Fill-O-Matic Speed	*Weight Variation *Friability *Hardness *Thickness *Disintegration
Packing Automatic packing line	* Packing line speed	*Attribute characteristics Critical, Major, Minor

Addition and Mixing

Check the dispersion of iron oxide red in purified water and mixing time and addition of Tween 80, for the validation batches.

Batch No	Date	Dispersion of Iron Oxide	Mixing Time

Check the quantity of granulating solution after preparation for the validation batches.

Batch No	Date	Theoretical QTY of Granulating Solution	Actual QTY of Granulating Solution

Sieving and Dry Mixing

Check the process step of Avicel PH 101, Premojel, and Lactose monohydrate were passed through a 500-micron sieve using the Russell sifter and then transferred to a polyethylene bag for blending for the validation batches

Batch No	Date	Sieve Size	Observation

Check the process step of Glimepiride and Lactose monohydrate were mixed manually in polyethylene bag and passed through 500-micron sieve using Russell sifter and transferred to polyethylene bag for blending for the validation batches

Batch No	Date	Sieve Size	Observation

Check the process step of Iron oxide red and Lactose monohydrate mixing manually in polyethylene bag and passing through 500-micron sieve using Russell sifter and transferring to polyethylene bag for blending for three validation batches.

Batch No	Date	Sieve Size	Observation

Check the process step of manual mixing in polyethylene bag for 3 minutes for Avicel PH 101, Premojel, and Lactose monohydrate for the three validation batches.

Batch No	Date	Time Limit 3 minutes	Observation

Wet Granulation

Check the following process step of wet granulation.

1. The loading of the manually pre mixed material in Diosna granulator/FBD from the polyethylene bag.
2. Mixing at speed 1 for 1 minute.
3. Addition of granulating solution through a sample inlet and again mixing with speed 1 for 1 minute. Adding of additional quantity of water for the validation batches

Batch No	Date	Loading of Material in Granulator Observation	1st Mixing Time 1 minute	2nd Mixing Time 1 minute	3rd Mixing Time 1 minute	Additional Water QTY

Drying of Granules and Loss on Drying

Check the process step drying of granules in a Lytzen Dryer/Aeromatic Flow bed dryer at 55°C for first 3 hours, scraping the semi-dried material and further drying for 3 hours at 55°C, further drying to obtain the target LOD, for the validation batches.

Batch No	Date	1st Drying	2nd Drying	Additional Drying

LOD Checking

Check the LOD at 80°C (limits 0.7%–1.0%) for the validation batches.

Batch No	Date	LOD Temperature	LOD Limits	Results

Check the process step grinding of dried granules and passing through a 1.0 mm sieve in Frewitt Granulator at medium speed and collecting of granules in polyethylene bag for the validation batches.

Batch No	Date	Sieve Size	Observation

Lubrication

Check the process step of Primojel, Kollidon CL, and Magnesium Stearate sifting through 500-micron SS sieve and finally mixing for 5 minutes manually for the validation batches

Batch No	Date	Sieve Size	Observation

Checked the final yield calculation of Granules for the validation batches.

Batch No	Date	Theoretical Batch weight	Actual Batch weight	% Yield

Compression of Tablets

Check the process step of compression using the Rotary compression machine for validation batches.

Test	Specification	Results		
Batch No's				
Weight of tablet	xxx mg ± 3.0%			
Thickness	2.4 ± 3.0 mm			
Hardness	NLT 7–13 kp			
Disintegration time	NMT 10 minutes			

The numbers given in the specifications are imaginary based on any given tablet and are only to provide an idea to the readers.

Uniformity of Weight

Weight variation Batch #

Target weight: xxx mg
Lower Limit weight: xxx mg
Upper Limit weight: xxx mg

Time								
Weight	1							
	2							
	3							
	4							
	5							
	6							
	7							
	8							
	9							
	10							
	Max							
	Min							
	Average							
	Range							
	S.D.							
	RSD							

CRITICAL PROCESS PARAMETERS (PACKAGING)

Blistering

During blistering in-process checks will be carried out for empty pocket, empty blister strip, improper sealing, leak test (10 blister), and defective tablets randomly on 32 blisters for the validation batches and results will be recorded in batch documents.

Batch No	Date	Empty Pocket	Empty Blister	Improper Sealing	Leak Test	Defective Tablets

FINAL PACKING RECORD (INSPECTION)

Critical, Major, and Minor Defects (SOP 123)

Attribute	AQL	Acceptance Criteria	Sample Size Cumulative (Code)	Total Rejected Samples		
				Batches Numbers		
Critical Defects (AQL 0)						
Wrong commodity	0.0%					
Wrong Batch No	0.0%					
Wrong Expiry	0.0%					
Wrong Manufacturing	0.0%					
Wrong Special stamping	0.0%					
Incorrect No of unites in C. Carton	0.0%					
Foreign Material	0.0%					
Empty unites	0.0%					
Major Defects (AQL 1.0%)						
Missing component	1.0%					
Illegible B.No	1.0%					
Illegible Mfg. Date	1.0%					
Illegible Exp. Date	1.0%					
Other						
Minor Defects (AQL 2.5%)						
Loose or torn labels/Carton/Box	2.5%					
Double labels	2.5%					
Printing Smeared Partly	2.5%					
Crooked label	2.5%					
Excessive Shrinkage label	2.5%					
others						
TOTAL						
Pass/Fail						
Conducted by				**Date:**		
Reviewed by				**Date:**		

SUMMARY OF RESULTS

Results Data Table

Add rows in the table below as per the number of samples which would depend on the size of a blender or container, the samples are collected from.

Tablet API – Assay **Specification: [a]1.0 mg/tablet (Limit:90.0%–105.0% of L.A)**			
Batch N°	**Assay-1.0 mg/tablet**	**% Assay Value**	
Maximum			
Minimum			
Range			
Average			
S.D			
RSD			
Acceptance Criteria	All individual values within 90.0%–105.0% of theoretical value of active ingredient and RSD for samples should be Less than 6%.		
Comments:			
Pass/Fail			
Reference:			
Conducted by		**Date**	
Reviewed by QC Manager		**Date**	

[a] The strength is imaginary and only to provide an idea to the readers.

9.2 Results Data Table

Related Substances **Specification:** **a. Tablet API Related Compound B: NMT 2.5%** **b. Any other individual impurity: NMT 0.5%** **c. Total impurities excluding Tablet API Related Compound B: NMT 1.0%** d. **Over all Total impurities: NMT 3.5%**				
Batch N°	**a)**	**b)**	**c)**	**d)**
Maximum				
Minimum				
Range				

(Continued)

Acceptance Criteria	All individual value for validation Batches samples should be within specification.		
Comments:			
Pass/Fail			
Reference:			
Conducted by		**Date**	
Reviewed by QA Manager		**Date**	

RESULT DATA TABLE (MICROBIAL ANALYSIS)

Specification: Microbial Limits **1 TAMC: NMT 100 cfu/gm** **2 Fungi: NMT 10 cfu/gm** **Pathogen: Absence of *Pseudomonas aeruginosa*** **Absence of *Staphylococcus aureus***					
	Results				
Batch No:	**TAMC**	**Fungi**	**Absence of *E. coli***	**Absence of *Pseudomonas aeruginosa***	**Absence of *Staphylococcus aureus***
Acceptance Criteria	Total aerobic microbial count should be LT 100 cfu/g, Fungi LT 10 cfu/g Pathogen: Absence of *E. coli*, *Pseudomonas aeruginosa*, and *Staphylococcus aureus*				
Comments:					
Pass/Fail					
Reference:					
Conducted by			**Date**		
Reviewed by QC Manager			**Date**		

FINAL BATCH YIELD

Check the final batch yield for the following validation batches.

Batch No	Batch Size	Target wt. of Tablets Limit: xxx mg	Theoretical Number of Tablets per Batch	Actual Number of Tablets per Batch	Yield

RISK ANALYSIS

Critical Process Steps	Possible Risk	Critical/Not Critical	Kind and Content of Control
During Manufacturing			
Blending Time	Unevenness	Critical	Visual display
Blending Speed	Unevenness	Critical	Visual display
Compression	Defectives (Chipping, Capping Sticking, etc.)	Critical	Visual
Manufacturing yield	Less productivity	Not Critical	Batch documents
Determination of assay	Inferior or higher content of drug substance	Critical	Assay
During Blistering and Packing			
Blistering/Box Packing	Less count May cause complaint	Critical	In-process checks every 1 hour
Critical, Major, Minor defects	May cause complaint	Critical	In-process checks every 1 hour
Printing of batch no. mfg. and expiry date	May cause complaint	Critical	In-process checks every 1 hour

OBSERVATION

Observation should be recorded during the validation process of the mentioned product

FINAL REPORT

Validation department will complete and review the final Validation report, declaring the successful completion of validation of product name. This final report will be reviewed and approved by the heads or manager of Validation and Production and the director of Quality Assurance as a minimum.

This section deals with the conclusion of the validation results and the discussion. This section also contains the recommendations of the validation committee.

7.2 Sampling and Testing Plan

In the process validation, sampling and testing plans must be developed to consider the specific attributes being measured and the risks associated with accepting a defective lot.

A sample is a portion of a material collected according to a defined sampling procedure. The size of a sample should be sufficient to allow all anticipated test procedures to be carried out including all repetitions and retention samples.

The Validation Protocol must have predetermined sampling plans and methods. The sampling procedures documented in the protocol should provide sample locations, sample size, and sample frequency and should also include the tests or assessments required. Samples must be representative of the population and the sampling plan should be designed in a way so that any segregation during blending and mixing, transport or handling could be determined. Besides, the sampling plan may also be able to assess any variability or other factors that might impact production consistency. The need to show the homogeneity of intermediates should be considered on a case-by-case basis depending on how the intermediate is used in subsequent processing. In general, studying the homogeneity of an intermediate is of less importance than that of a final product, especially if the intermediate will be further processed in the next or final step of the process. However, in case the intermediate is critical to the quality of the final product prepared from it, ensuring that the blend homogeneity of the intermediate must be considered in the sampling plan.

SAMPLING LOCATIONS

Equipment structure, shape, and dimensions should be taken into consideration when establishing sampling locations. Blend uniformity samples should be collected directly from the blender immediately prior to discharge. It is generally considered that a blend uniformity sampling from tumble blending should be carried out from, at least ten locations in the blender. These locations should be chosen to represent potential areas of poor blending. At minimum samples should be selected from at least two depths along the axis of the blender.

Similarly, for ribbon blending, a minimum of 20 locations in the blender should be sampled and chosen to adequately address probable dead spots especially in the corners and discharge areas. Samples may also be taken from discharge stream or storage containers like drums and bins in case it is not possible to collect samples from the blender. At least three replicate samples will be taken from each location with one samples tested per location.

Blend uniformity acceptance criteria: RSD = 5.0% and all individuals are within the mean ± 10% (absolute).

DOI: 10.1201/9781003366393-9

FINISHED PRODUCT SAMPLING

The finished product sampling quantity and sample numbers for each validation study should be based on risk. For example, a new formulation may carry significantly higher risks to the patient. In such a case, the following rules should be considered at minimum for filling/packaging sampling:

- Beginning of Filling: Logically the first ten samples should be collected and tested as "Beginning samples".
- Middle: During the entire filling process, samples should be collected at a frequency established in the protocol as per the process run and then ten samples collected from the entire pool to create the "Middle samples".
- End: Like the first ten packages, the last ten packages just before the end of filling should be collected as the "End samples".

In addition, if there are critical incidences during the packaging run which may affect the product, then sampling at start-up immediately following such an event should also be considered. The rationale behind collection and/or testing of these samples should be outlined in the protocol.

Sampling after a Change in Raw Material Source

A change in the supplier or manufacturer of a raw material is critical and must be assessed for the impact on product's quality. Based on the assessment, decision is made if the change (made to an active or excipient in a formulation) requires revalidation. The sampling requirement for the revalidation will be based on risk and documented in the Raw Material Evaluation Report.

For the products where routine blend testing is not performed, there must be an impact assessment made on blend uniformity and the requirements of the blend specification.

It may be a good idea to conduct additional monitoring and testing during or post validation batches of those products. This should also be documented in the validation protocol.

PROCESS AND EQUIPMENT CHANGE

A similar strategy, as mentioned above, needs to be carried out in case of a change in the process and/or equipment. Each process or equipment change should be addressed individually and the validation requirements must be documented in the protocol.

Sampling and testing should be based on risk. For those products where routine blend testing is not performed, an assessment of the impact on blend uniformity and the requirements of the blend specification should be performed. Additional monitoring and testing during or post validation batches may be required.

Specification testing of the extremes of the spread will allow full characterization of the content uniformity. Batch release testing will be performed as per specification test requirements and must be included in the validation report. A validation batch using QC reductive testing must meet all QC reductive test requirements for the validation samples and specification requirements for the release of sale samples.

SAMPLING AND TESTING PLAN TO PERFORM ACCEPTABLE QUALITY LIMIT

AQL (Acceptable Quality Limit) Sampling is a method widely used to define a production order sample to find whether or not the entire product order has met the client's specifications. Based on the sampling data, the customer can make an informed decision to accept or reject the lot. There are various sampling plans such as single sample plan or multiple sampling plans for performing AQL.

AQL sampling is performed to identify and accept numbers of defective units (i.e. critical, major, and minor defects) in a lot or a batch.

- AQL sampling can be performed at stages such as but not limited to after compression, during online compression, after coating, after filling of capsules, during the online filling of capsules.
- Samples shall be withdrawn as per AQL chart.
- During online processing of a batch, sample size at a particular frequency can be calculated as per the below mentioned formula:

$$n = \frac{N}{R} \times F$$

Where, n = sample size at each interval
N = Total No. of samples
R = Total run time of machine in minutes
F = Frequency or time interval in minutes

- Sampling can also be performed after processing of a batch by dividing the total sample size with the total nos. of containers containing the batch. Sampling shall be performed by taking equal nos. of samples from individual containers randomly.

CONCLUSION

Sampling performed in various stages of solid oral dosage formulation manufacturing industries plays a critical role in lot or batch acceptance investigation in case of any deviation or incidence which in turn helps to maintain or improve or develop the quality of a product according to regulatory or in house requirements of a pharmaceutical industry in view of patient safety.

Sampling methods other than those mentioned above can be adopted having sound scientific justification and rationale. There may be many more sampling errors associated with sampling other than those mentioned in the article. Sampling errors should be identified and can be avoided by regular training and monitoring of samplers and/or by performing sampler validation in sampling activities.

Sampling and testing plans for every stage of the process for a liquid dosage form are defined in the chart below.

SAMPLING AND TESING PLAN FOR PRODUCT ABC

LOT:__________ W/O__________ DATE:__________ RUN No.__________

In-Process

Sample No.	Nature of Sample	Location/Time	Responsibility	Test to Be Preformed	Sampled By	Time/Date of Sample
	In-Process	Sample Port (100 mL)	Validation	IP (Write the tests to be performed for all respective areas)		
	In-Process	Sample Port (100 mL)	Validation	IP		
	Bulk Hold (When required)	Sample Port – Time 0	Validation	FP		
		Sample Port – Time 24 hours	Validation	FP		

Condensate Purge

Sample No.	Nature of Sample	Location/Time	Responsibility	Test to Be Preformed	Sampled By	Time/Date of Sample
	Condensate Purge	XX L (100 mL)	Validation	Test Active		
		XX L (100 mL)	Validation	Test Active		
	(Continue as required)	XX L (100 mL)	Validation	Test Active		

Batch Homogeneity

Sample No.	Nature of Sample	Location/Time	Responibility	Test to Be Preformed	Sampled By	Time/Date of Sample
	Batch Homogeneity	Beginning of Fill (XX bottles)	PKG/Validation	Test Active		
		Beginning of Filling (XX bottles)	PKG/Validation	Test Active		
		Middle of Filling (XX bottles)	PKG/Validation	Test Active		
		Middle of Filling (XX bottles)	PKG/Validation	Test Active		
		End of Filling (XX bottles)	PKG/Validation	Test Active		
		End of Filling (XX bottles)	PKG/Validation	Test Active		

QUESTIONS AND ANSWERS

1. **Why a sampling Plan in validation is important?**

   Answer: In the process validation, sampling and testing plans must be developed to consider the specific attributes being measured and the risks associated with accepting a defective lot.

2. **Define a sample**

   Answer: A sample is a portion of a material collected according to a defined sampling procedure.

3. **What are the significant aspects of a sampling procedure?**

   Answer: The sampling procedures documented in the protocol should provide;

   a. Sample locations, sample size, and sample frequency and should also include the tests or assessments required.
   b. Must be representative of the population.
   c. Should be designed in a way so that any segregation during blending and mixing, transport, or handling could be determined.
   d. Should also be able to assess any variability or other factors that might impact production consistency.

4. **How many sampling locations are recommended from a blender to assess the blend uniformity?**

   Answer: It is generally considered that a blend uniformity sampling from tumble blending should be carried out from, at least ten locations in the blender.

   Whereas for ribbon blending, a minimum of 20 locations in the blender should be sampled and chosen to adequately address probable dead spots especially in the corners and discharge areas.

5. **What should be the logical approach in choosing the Beginning, Middle, and End samples during a filling process?**

   Answer: During the filling process, the following rules should be considered at minimum for filling/packaging sampling to establish the Beginning, Middle, and End samples:

   - Beginning of Filling: Logically the first ten samples should be collected and tested as "Beginning samples".
   - Middle: During the entire filling process, samples should be collected at a frequency established in the protocol as per the process run and then ten samples collected from the entire pool to create the "Middle samples".
   - End: Like the first ten packages, the last ten packages just before the end of filling should be collected as the "End samples".

6. What other changes in a process may trigger a re-validation and consequently change in sampling plan would be required.

Answer: The following changes may trigger a re-validation and a changed sampling plan;

a. A change in the supplier or manufacturer of a raw material is critical and must be assessed for the impact on product's quality. Based on the assessment, decision is made if the change (made to an active or excipient in a formulation) requires revalidation. The sampling requirement for the revalidation will be based on risk and documented in the Raw Material Evaluation Report.

b. A similar strategy, as mentioned above, needs to be carried out in the case of a change in the process and/or equipment. Each process or equipment change should be addressed individually and the sampling and testing should be based on risk.

7.3 Process Validation Report Writing

The process validation report is a document which demonstrates evidence that a manufacturing process is capable of consistently delivering quality products. It provides a proper documentation of process validation protocols. Process validation reports are generally completed before the routine commercial production for a new product. The validation report may clarify if the batches discussed in the report are going to be used for commercial purpose or not. Pharmaceutical manufacturing industries typically use a process validation report template to ensure compliance with US Food and Drug Administration (FDA) requirements.

The validation report should contain an executive summary listing the most prominent results of the validation. A short introduction stating who produced it, when, and for what purpose. The process validation report generally includes the details of the materials used with the batch numbers, the equipment with the ID numbers, and the personnel who carried out the processes with their training records ensuring that the process controls were performed by experienced operators.

The raw data is further discussed in the report and if there has been any anomaly during the validation study, it is addressed in the report with the investigation details as well as the outcome of the investigation and corrective and preventative actions taken.

In the following pages, a validation report is being given step by step to clarify how a process validation report is written.

OBJECTIVE

The objective of the report is to document the process results and process parameters obtained during the manufacture of [enter blend/granulate name and batch number] and [enter Capsule/Tablet product name and batch number] in the Manufacturing facility at [enter site name].

The [enter blend/granulate name and batch number] are the [first/second/third] validation blend in this study manufactured for the Local market. The [enter Capsule/Tablet product name and batch number] are the [first/second/third] run of three validation batches to be manufactured for the [Local /export] markets. Refer to table 1.0 below for details on all the validation runs covered in this report and from previous interim reports.

VALIDATION STRATEGY

The manufacturing process of the xyz blend name (commercial lot size Y kg) and [enter product caps/tab] (enter batch size or commercial batch size may depend

DOI: 10.1201/9781003366393-10

on market demand) were validated under the control of the Technical Services Department. [Enter for both export and Local if applicable].

A [concurrent/prospective/retrospective] validation approach was used for this validation study.

Justify concurrent e.g., because some of the [enter product name] dosages are not manufactured very frequently in the Manufacturing facility at [enter site name].

All validation batches were manufactured following the same manufacturing process as detailed in the manufacturing instructions [give Manufacturing Instruction number and current versus protocol version number/date].

All critical process variables in the manufacturing process [blending/granulation and Tableting /encapsulation] were reviewed. [These/this] validation [batch/es] were analyzed as set out in [enter section reference] of the protocol [enter protocol number] and the results are summarized in [enter Section reference] of this report.

This report will document the number of blend / granulate samples from the run, e.g., minimum of ten locations from the blender to show blend uniformity/homogeneity [dead spot] characterization in modified processes only, reference [enter no] and Blend sampling strategy report [enter no]. NIR blend analysis will be run on all blend samples to assess blend uniformity.

REPORT APPROVALS SHEET

PRODUCT NAME:	
CODE:	
BATCH NUMBER:	
PRODUCTION STAGE:	

Table of Content

Sr.No.	Description	Page No.
	REPORT APROVALS SHEET	1
	EXECUTIVE SUMMARY	4
	INTRODUCTION	5
	TYPE OF VALIDATION	5
	VALIDATION STRATEGY	6
	DOCUMENTATION	7
	CRITICAL PROCESS PARAMETERS (MANUFACTURING)	8
	Formulation	8
	Equipment used	9
	Manufacturing process	10
	Process flow, variables and responses	10
	Addition and mixing	11
	Addition of Iron oxide red, etc.	11

(*Continued*)

Table of Content		
Sr.No.	Description	Page No.
	Quantity of Granulating Solution	11
	Mixing and Sifting of Avicel 101, etc.	11
	Mixing and Sifting of API	11
	Mixing and Sifting of Iron oxide red, etc.	12
	Mixing of all Sifted material	12
	Preparation of wet granules	12
	Drying of Granules	13
	Checking of LOD	13
	Grinding of Dry Granules	13
	Lubrication	14
	Yield Calculation of Granules	14
	Compression	14
	Uniformity of weight	15–17
	CRITICAL PROCESS PARAMETERS (PACKAGING)	18
	Blistering defects	18
	Final Packaging Record	19
	SUMMARY OF QC RESULTS	20
	Results data table of API assay	20
	Results data table for Impurities	22
	Results data table of Microbial Analysis	23
	BATCH YEILD	24
	RISK ANALYSIS	25
	COMPLIANCE TO FINISHED PRODUCT SPECIFICATION	26–27
	OBSERVATION	28
	CONCLUSION	29
	REVALIDATION	30

2 EXECUTIVE SUMMARY

This process validation report has the compilation of data, including manufacturing process variables, packaging variables, and analytical results for the product xyz 1 mg Tablets which was successfully manufactured at the plant ABC Pharmaceutical industries.

Three consecutive validation batches were manufactured of xyz 1 mg Tablets and the complete batch document were reviewed, and related documents like Batch Manufacturing formula, Batch Packaging record, QC analytical methods, and related SOPs were checked.

In order to achieve the goal of process validation the analytical data of three consecutive batches was compiled and statistically evaluated and represented graphically for the assay of active ingredients.

All three batches meet the acceptance criteria for all quality attributes set up in the finished product specification and related SOPs.

3 INTRODUCTION

3.1 Objective of the Validation

The aim of this report is to document that the process of production of three consecutive batches of pharmaceutical product xyz 1 mg Tablets is effective in consistency and reproducibility and meets all predetermined specifications and quality attributes using the actual process to verify at least the following:

- Final Mixing homogeneity
- Physical parameters
- Content of Active ingredient
- Normal operating parameters
- Operating procedures for product manufacturing

The validation report was focused to the critical points for quality assurance of the production process in the way that:

- Content of active ingredient – assay
- Microbial Analysis

Complying with specification and acceptance criteria established in finished product specification for xyz 1 mg Tablets.

This report presents the results of the following two consecutive production batches for xyz 1 mg Tablets.

Batch No.	Batch Size	Theoretical Units Produced	Manufacturing Date	Expiry Date
001	xx kg	100,000		
002	xx kg	100,000		
003	xx kg	100,000		

4 TYPE OF VALIDATION

Prospective validation

5 VALIDATION STRATEGY

Three consecutive batches of the above-mentioned product were manufactured in the manufacturing area. These batches were used for the above-mentioned type of process validation study. The work was focused on the evaluation of the following steps:

i. Review of the manufacturing formula and method, packaging instructions, raw material and finished product specifications and related SOPs.
ii. Issuance of Raw Materials and Equipment used during manufacturing.
iii. Mixing time and mixer speed at different manufacturing steps.
iv. Temperature of vessel during mixing.
v. The routine in-process control data and the critical parameters according to the manufacturing formula and method (MFM).
vi. Manufacturing and final batch yield.
vii. Critical packaging parameters.
viii. Physicals, Chemical, and microbiological properties of the active ingredients for finished product.

6 DOCUMENTATION

Check the following in use documents before the start of process validation.

	Reference No.	Issue Date
Manufacturing Formula and Method (MFM)		
Master Packaging Instructions		
Finished Product Specification		
In-process Specification (Manufacturing)		
In-process Specification (Packaging)		
Raw Material Specifications		
1: API (Micronized) 3: Lactose Monohydrate 4: Microcrystalline Cellulose (Avicel) 5: Sodium Starch Glycolate (Primojel) 6: Crospvidone (Kollidon CL) 7: Polysorbate 80 (Tween 80) 8: Iron Oxide Red 9: Iron Oxide Red 10: Magnesium Stereate 11: Purified Water		

7 CRITICAL PROCESS PARAMETERS (MANUFACTURING)

7.1 Formulation (Active/Inactive Ingredients)

Composition of Product

Scale mg/tablet	Item No.	Material's Name	Code	Quantity/Batch 14.0 kg (100,000 tablet)	QC Status
	1	API (Micronized)			Released
	4	Lactose Monohydrate			Released
	3	Microcrystalline Cellulose (Avicel pH101)			Released
	4	Sodium Starch Glycolate (Primojel)			Released
	5	Crospvidone (Kollidon CL)			Released
	6	Polysorbate 80 (Tween 80)			Released
	7	Iron Oxide Red			Released
	8	Magnesium Stearate			Released
	9	Purified water			Released

Names of raw materials in the table above are given as an example as most of the solid dosage forms are manufactured using the same raw materials as excipients. However, the main logic here is to only provide a basic idea to the readers.

7.2 Equipment Used

The details of major equipment used during manufacturing and packaging are given in below table.

Sr. No.	Equipment Name	Manufacturer's Name	Model	Serial No.
	Sifter			
	Mixer			
	Tray Dryer			
	Dry Granulator			
	Tablet Press			

The names of the equipment in the table above are the ones generally used in the manufacture of a solid dosage form product. However, their names are given here only to provide an idea to the readers.

7.3 Manufacturing Process

7.3.1 Process Flow, Variables and Responses for Tablet Products

Process Flow	Control Variables	Measured Responses
Addition of Raw Materials Active and Excipients	* Weighing / Sampling * Warehouse / QA	* Released by QC
↓		
Sizing Sifter Milling / Sieving	* Screen Type * Screen Size	*Granules Size Distribution
↓		
Mixing Dry and Wet mixing Addition of granulating fluid	* Mixing Time * Mixer Speed	
↓		
Drying Oven / Fluid Bed	*Drying Time *Drying Temperature	*LOD *Moisture Content
↓		
Blending Double Cone Blender / Tumbler Addition of Raw Materials Binders / Lubricants / Disintegrants	*Load Size *rpm *Blending Time	*Blend Uniformity *Flow Characteristics
↓		
Compressing High Speed Rotary Machine	*Compression Rate *Granules Feed Rate *Fill-O-Matic Speed	*Weight Variation *Friability *Hardness *Thickness *Disintegration
↓		
Packing Automatic packing line	* Packing line speed	*Attribute characteristics Critical, Major, Minor

7.3.2 Addition and Mixing

Sieving and Dry Mixing

7.3.2.1 Avicel pH 101, Primojel, and Lactose monohydrate passed through a 630 μm sieve using Russel sifter and transferred to a polyethylene bag for blending. This process step was checked in the batch document and was found satisfactory for the following three validation batches

Batch No.	Date	Sieve Size	Observation
001		630 μm	Satisfactory
002		630 μm	Satisfactory
003		630 μm	Satisfactory

7.3.2.2 API and Lactose monohydrate were mixed manually in polyethylene bag and passed through 630 μm sieve using sifter and transferred to polyethylene bag for blending.

This process step was checked in the batch document and was found to be satisfactory for the following three validation batches.

Batch No.	Date	Sieve Size	Observation
001		630 μm	Satisfactory
002		630 μm	Satisfactory
003		630 μm	Satisfactory

7.3.2.3 Iron oxide Red and Lactose monohydrate were mixed manually in a polyethylene bag and passed through a 250 μm sieve using Russel sifter and transferred to Diosna for mixing. This process step was checked in the batch document and was found to be satisfactory for the following three validation batches

Batch No.	Date	Sieve Size	Observation
001		250 μm	Satisfactory
002		250 μm	Satisfactory
003		250 μm	Satisfactory

Preparation of Granulating Solution

7.3.2.4 The mixing of iron oxide red in purified water and the addition of Tween 80 were reviewed and recorded from the MFM for the validation batches.

Batch No.	Date	Dispersion of Iron Oxide	Mixing Time
001		uniform	5 minutes
002		uniform	5 minutes
003		uniform	5 minutes

7.3.2.5 The quantity of granulating solution was reviewed and recorded from the MFM for the two validation batches.

Batch No.	Date	Theoretical QTY of Granulating Solution	Actual QTY of Granulating Solution
001		x.00 kg	3.03 kg
002		x.00 kg	3.03 kg
003		x.00 kg	3.03 kg

7.3.2.6 The manual mixing in polyethylene bag for 3 minutes was done for Avicel pH 101, Primojel, and Lactose monohydrate. This process step was checked in the batch document and was found to be satisfactory for the following validation batches

Batch No.	Date	Time Limit 3 Minutes	Observation
001		500 μm	Satisfactory
002		500 μm	Satisfactory
003		500 μm	Satisfactory

Wet Granulation

7.3.2.7 The process of wet granulation was checked in the batch document for the following step

1. The loading of manually pre-mixed material in the granulator from the polyethylene bag.
2. Mixing at speed 1 for 1 minute.
3. Addition of granulating solution through a sample inlet and again mixing with speed 1 for 1 minute. Adding of additional quantity of water. All the steps were found to be satisfactory for the following validation batches.

Batch No.	Date	Loading of Material in Granulator	1st Mixing Time 1 Minute	2nd Mixing Time 1 Minute	3rd Mixing Time 1 Minute	Additional Water QTY
001		Satisfactory	1 minute	1 minute	1 minute	x.0kg
002		Satisfactory	1 minute	1 minute	1 minute	x.0kg
003		Satisfactory	1 minute	1 minute	1 minute	x.0kg

Drying of Granules and LOD

7.3.2.8 The Drying of granules in a Dryer was done at 55°C for 10 hours, was checked in the batch document, and was found to be satisfactory for the following validation batches.

Batch No.	Date	Temperature	Drying
001		55°C	xx Hours
002		55°C	xx Hours
003		55°C	xx Hours

LOD Checking

7.3.2.9 The LOD was noted at 80°C (limits 0.7%–1.0%) checked in the batch document and was found to be satisfactory for the validation batches.

Batch No.	Date	LOD Temperature	LOD Limits	Results
001		80°C	0.7%–1.0%	0.90%
002		80°C	0.7%–1.0%	0.80%
003		80°C	0.7%–1.0%	0.90%

7.3.2.10 Grinding of Dried granules and passing through a 1.0mm sieve in a Granulator at medium speed and collecting of granules in SS drum was checked in the batch document and was found to be satisfactory for the validation batches.

Batch No.	Date	Sieve Size	Results
001		1.0mm	Satisfactory
002		1.0mm	Satisfactory
003		1.0mm	Satisfactory

Numbers given in the tables above for measuring time or any other parameter are not referring to any specific product but are only to provide an idea to the readers.

Lubrication

7.3.2.11 Primojel, Kollidon CL, and Magnesium Stearate were sifted through a 500μm SS sieve and added to the ground and sieved granules, and finally mixed for 5 minutes manually; this process was reviewed and recorded from the MFM for the two validation batches.

Batch No.	Date	Sieve Size	Observation
001		500μm	Satisfactory
002		500μm	Satisfactory
003		500μm	Satisfactory

7.3.2.12 The final yield calculation of Granules for the three batches re-checked from batch document, and found satisfactory.

Batch No.	Date	Theoretical Batch Weight	Actual Batch Weight	% Yield
001		14.0kg	13.86kg	99.0%
002		14.0kg	13.88kg	99.14%
003		14.0kg	13.7kg	97.85%

Compression of Tablets

7.3.2.13 Tablets were compressed using Rotary compression machine. This process step was checked in the batch document and the following parameters were found satisfactory for the following two validation batches.

Test	Specification	Results		
Batch Nos		001	002	003
Weight of tablet	140 mg ± 3.0%	140 mg	140 mg	140 mg
Thickness	2.4 ± 3.0 mm	2.4 mm	2.5 mm	2.5 mm
Hardness	NLT 7–13 kp	9 kp	9 kp	10 kp
Disintegration Time	NMT 10 minute	7 minutes	6 minutes	0.5 minute

In-Process Testing – Use What Is Applicable

The analysis performed throughout this run is ten tablets at the following time points:

- The first ten tablets from the start of the compression process
- From a composite* sample of 20 dosages from each bucket throughout the run
- Critical Time Points – first ten tablets off the machine, after the event.

Examples of Critical Time Points

- Major break down of tablet machine
- After an extended break in compression (i.e. outside normal operational stops and starts) and blockages
- The last ten tablets off the machine

The potency of the unit dosage with both the highest and lowest absorbance from the NIR scans over the entire run was analyzed for potency by the QC Lab to equate the absorbance to the actual concentration.

In-process weight results and statistical analysis performed to identify the state of process control are documented in this report in Section 7.3.3.

This report documents the sampling and testing required and performed on the finished product.

Full product testing on the routine end – of the – run sample of tablets was required apart from Dissolution, Assay, and Content Uniformity which were required to be tested on the additional validation samples. The additional validation samples, taken from seven different time points distributed throughout the validation run, include start, middle, and end of run. Assay and Content Uniformity testing were required on 70 tablets (ten tablets from each time point) and [Comparative] Dissolution testing was required on 12 [or six if only dissolution profile] tablets from the end – of the – run sample. The Comparative Dissolution profile is required to be performed on three batches of the current process and compared to each validation batch [if applicable].

7.3.3 Uniformity of Weight

Weight variation Batch # 001

Target weight: 140 mg
Lower Limit weight: 133 mg
Upper Limit weight: 147 mg

	Time	16.00	16.30	17.00	17.30	18.00	18.30	19.00	19.30	14.00
Weight	1	143	140	139	140	139	140	138	136	138
	2	138	144	144	142	143	145	143	141	142
	3	140	142	142	141	142	136	134	134	147
	4	139	138	142	143	142	149	147	145	140
	5	141	141	140	140	146	139	137	135	145
	6	139	139	144	142	141	142	140	138	139
	7	145	143	141	137	144	144	142	140	143
	8	139	140	140	138	140	139	137	135	140
	9	141	140	138	140	143	142	140	138	138
	10	144	145	140	137	143	140	138	136	140
	Max	145	145	144	143	146	149	147	145	147
	Min	138	138	138	137	139	136	134	134	138
	Average	141	141	141	140	142	142	140	138	141
	Range	7	7	6	6	7	13	13	11	9
	S.D	2.24	2.25	2.00	2.11	2.00	3.69	3.69	3.54	3.01
	RSD	1.59%	1.59%	1.42%	1.51%	1.41%	2.60%	2.64%	2.57%	2.13%

All the weights given in the target, lower and upper limits are not for any specific product but are only an example to present an idea to the readers.

Weight variation Batch # 002

Target weight: 140 mg
Lower Limit weight: 133 mg
Upper Limit weight: 147 mg

	Time	9.30	10.00	10.30	11.00	11.30	12.45	13.00	13.30	14.00	14.30
Weight	1	139	138	140	137	139	137	142	140	137	138
	2	137	141	143	141	143	135	146	143	141	144
	3	140	139	141	143	145	140	139	139	137	141
	4	145	142	144	140	142	143	143	137	139	141
	5	139	139	141	135	137	137	145	140	143	140
	6	135	141	143	142	144	133	140	146	142	143
	7	142	138	140	134	136	140	141	143	141	138
	8	145	140	142	138	140	143	144	142	146	140
	9	140	138	140	142	144	138	142	140	142	142
	10	140	144	146	146	148	136	140	138	139	136
	Max	145	144	146	146	148	143	146	146	146	144
	Min	135	138	140	134	136	133	139	137	137	136
	Average	140	140	142	140	142	138	142	141	141	140
	Range	10	6	6	12	12	10	7	9	9	8
	S.D	3.16	2.00	2.00	3.77	3.77	3.29	2.30	2.70	2.79	2.45
	RSD	2.25%	1.43%	1.41%	2.69%	2.66%	2.38%	1.62%	1.92%	1.98%	1.75%

All the weights given in the target, lower and upper limits are not for any specific product but are only an example to present an idea to the readers.

Weight variation Batch # 003

Target weight: 140 mg
Lower Limit weight: 133 mg
Upper Limit weight: 147 mg

	Time	9.00	10.00	11.00	12.00	13.00	14.00
Weight	1	140	140	140	141	143	141
	2	143	139	138	143	142	142
	3	143	142	139	141	145	142
	4	143	138	141	140	142	140
	5	142	141	137	144	144	144
	6	143	141	140	142	142	141
	7	143	143	142	141	141	140
	8	142	139	140	142	144	139
	9	142	138	143	141	142	144
	10	142	139	143	140	140	141
	Max	143	143	143	144	145	144
	Min	140	138	137	140	140	139
	Average	142	140	140	142	143	141
	Range	3	5	6	4	5	5
	S.D	0.95	1.70	2.00	1.27	1.51	1.65
	RSD	0.67%	1.21%	1.43%	0.90%	1.06%	1.16%

All the weights given in the target, lower and upper limits are not for any specific product but are only an example to present an idea to the readers.

In the actual report, the set of ten tablets can be repeated as per the pre-approved protocol for each of the validation batches. Depending on the size of the batches and duration of compression, the set of ten tablets for weight variation could vary from three to five or even more if needed and the results can be tabulated as per the tables shown above.

Similar tables could be used to tabulate the results of assay, dissolution, and content uniformity tests as described under the section "in-process testing".

The same approach could be used for liquid dosage forms validation batch reports. The physical attributes of tablets and capsules could be replaced with pH, specific gravity, viscosity, etc.

8 CRITICAL PROCESS PARAMETERS (PACKAGING) (SOP QAS-0110)

8.1 Blistering

During blistering in-process checks were carried out for empty pocket, empty blister strip, improper sealing, leak test (ten blister), and defective tablets were checked randomly on 32 blisters for the batches and results were found to be satisfactory. A detailed report is available in batch file.

Batch No.	Date	Empty pocket	Empty blister	Improper sealing	Leak test	Defective tablets
001		0/32	0/32	0/32	0/10	0/32
002		0/32	0/32	0/32	0/10	0/32
003		0/32	0/32	0/32	0/10	0/32

Numbers of defects given in the table above are only examples to provide an idea to the readers.

8.2 Final Packing Record (Inspection)

Critical, Major, and Minor Defects (SOP 123)

Attribute	AQL	Acceptance Criteria	Sample Size Cumulative 160 (k)	Total Rejected samples		
				Batches Numbers		
				001	002	003
Critical Defects (AQL 0)						
Wrong commodity	0.0%	0	160–192			
Wrong Batch No	0.0%	0	160–192			
Wrong Expiry	0.0%	0	160–192			
Wrong Manufacturing	0.0%	0	160–192			
Wrong Special stamping	0.0%	0	160–192			
Incorrect No of unites in C. Carton	0.0%	0	160–192			
Foreign Material	0.0%	0	160–192			
Empty unites	0.0%	0	160–192			
Major Defects (AQL 1.0%)						
Missing component	1.0%	4	160–192			
Illegible Batch No.	1.0%	4	160–192			
Illegible Mfg. Date	1.0%	4	160–192			
Illegible Exp. Date	1.0%	4	160–192			
Other						
Minor Defects (AQL 2.5%)						
Loose or torn labels / Carton/Box	2.5%	8	160–192			
Double labels	2.5%	8	160–192			
Printing Smeared Partly	2.5%	8	160–192			
Crooked label	2.5%	8	160–192	0 0	0 0	0 0
Excessive Shrinkage label	2.5%	8	160–192	0 0	0 0	0 0
others						
TOTAL				0	0	0
Pass/Fail				**PASS**	**PASS**	**PASS**
Compiled and Reviewed by				**Date:**		

9 SUMMARY OF RESULTS

9.1 Result Data Table

<table>
<tr><th colspan="4">API – Assay
Specification: 1.0 mg/tablet (Limit:90.0%–105.0% of LA)</th></tr>
<tr><th>Batch No.</th><th>Assay-1.0 mg/tablet</th><th colspan="2">% Assay value</th></tr>
<tr><td>001</td><td></td><td colspan="2"></td></tr>
<tr><td>002</td><td></td><td colspan="2"></td></tr>
<tr><td>003</td><td></td><td colspan="2"></td></tr>
<tr><td>Maximum</td><td></td><td colspan="2"></td></tr>
<tr><td>Minimum</td><td></td><td colspan="2"></td></tr>
<tr><td>Range</td><td></td><td colspan="2"></td></tr>
<tr><td>Average</td><td></td><td colspan="2"></td></tr>
<tr><td>S.D</td><td></td><td colspan="2"></td></tr>
<tr><td>RSD</td><td></td><td colspan="2"></td></tr>
<tr><td>Acceptance Criteria</td><td colspan="3">All individual values within 90.0%–105.0% of the theoretical value of active ingredient and RSD for 2 Batch samples should be less than 6%.</td></tr>
<tr><td colspan="4">Comments: All result meet Finished product specification and acceptance criteria.</td></tr>
<tr><td>Pass / Fail</td><td colspan="3">Pass</td></tr>
<tr><td>Compiled by</td><td></td><td>Date</td><td></td></tr>
<tr><td>Reviewed by QA Manager</td><td></td><td>Date</td><td></td></tr>
</table>

Dissolution Test

Target: 90%–110% mg

Lower Limit:

Upper Limit:

	Time	9.00	10.00	11.00	12.00	13.00	14.00
Assay	1						
	2						
	3						
	4						
	5						
	6						
	7						
	8						
	9						
	10						
	Max						
	Min						
	Average						
	Range						
	S.D						
	RSD						

Samples collected at the frequent interval will be tested for dissolution assay and results will be tabulated in the table above for all three validation batches. Any anomaly will be discussed and rationalized in the report if required.

9.2 Results Data Table

<table>
<tr><td colspan="6">Related Substances
Specification:
a. API: NMT 0.4%
b. API Urethane: NMT 0.1%
c. API-3-Isomer: NMT 0.2%
d. Any unspecified individual: NMT 0.1%
e. Total excluding Sulfonamide: NMT 0.2%</td></tr>
<tr><th>Batch No.</th><th>a.</th><th>b.</th><th>c.</th><th>d.</th><th>e.</th></tr>
<tr><td>002</td><td></td><td></td><td></td><td></td><td></td></tr>
<tr><td>003</td><td></td><td></td><td></td><td></td><td></td></tr>
<tr><td>004</td><td></td><td></td><td></td><td></td><td></td></tr>
<tr><td>Maximum</td><td></td><td></td><td></td><td></td><td></td></tr>
<tr><td>Minimum</td><td></td><td></td><td></td><td></td><td></td></tr>
<tr><td>Range</td><td></td><td></td><td></td><td></td><td></td></tr>
<tr><td>Acceptance Criteria</td><td colspan="5">All individual value for validation Batch samples should be within specification.</td></tr>
<tr><td colspan="6">Comments: All result meet Finished product specification and acceptance criteria.</td></tr>
<tr><td>Pass/Fail</td><td colspan="5">Pass</td></tr>
<tr><td>Compiled by</td><td colspan="2"></td><td>Date</td><td colspan="2"></td></tr>
<tr><td>Reviewed by
QA Manager</td><td colspan="2"></td><td>Date</td><td colspan="2"></td></tr>
</table>

Sulfonamide is one of the related substances that could be found in the products and is given here as an example only.

9.3 Result Data Table (Microbial Analysis)

Specification: Microbial Limits

1. TAMC: NMT 100 cfu/g
2. Fungi: NMT 10 cfu/g

Pathogen: Absence of Pseudomonas aeruginosa
Absence of Staphylococcus aureus

Batch No.	Results				
	TAMC	**Fungi**	**Absence of *E. coli***	**Absence of *Pseudomonas aeruginosa***	**Absence of *Staphylococcus aureus***
001	< 10 cfu/g	< 10 cfu/g	Absent	Absent	Absent
002	< 10 cfu/g	< 10 cfu/g	Absent	Absent	Absent
003	< 10 cfu/g	< 10 cfu/g	Absent	Absent	Absent

Acceptance Criteria	Total aerobic microbial count should be LT 100 cfu/g, Fungi LT 10 cfu/g Pathogen: Absence of *E. coli*, *Pseudomonas aeruginosa*, and *Staphylococcus aureus*		
Comments: All Results Conforms Finished product specification.			
Pass/Fail	**PASS**		
Compiled by		**Date**	
Reviewed by QA Manager		**Date**	

10 FINAL BATCH YIELD

Final batch yield was checked and calculated on the basis of BPR for the following two batches and found to be unsatisfactory the investigation report and other details available in the batch documents.

Batch No.	Batch Size	Target wt of Tablets Limit:.140 mg SOP 123	Theoretical Number of Tablets per batch	Actual Number of Tablets per batch	Yield
001		140.0 mg	100,000	97,000	97.0%
002		140.0 mg	100,000	97,500	97.5%
003		140.0 mg	100,000	95,700	95.70%

All the numbers given in the table above for target weights or number of tablets are not specific to any product but are presented as an example for the readers

11 RISK ANALYSIS

Critical Process Steps	Possible Risk	Critical/Not Critical	Kind and Content of Control
During Manufacturing			
Blending time	Unevenness	Critical	Visual display
Blending speed	Unevenness	Critical	Visual display
Compression	Defectives (Chipping, Capping Sticking, etc.)	Critical	Visual
Manufacturing yield	Less productivity	Not critical	Batch documents
Determination of assay	Inferior or higher content of drug substance	Critical	Assay
During Blistering and Packing			
Tube filling	Less count May cause complaint	Critical	In-process checks every 1 hour
Critical, Major, Minor defects	May cause complaint	Critical	In-process checks every 1 hour
Printing of batch no. mfg. and expiry date	May cause complaint	Critical	In-process checks every 1 hour

12 COMPLIANCE TO FINISHED PRODUCT SPECIFICATION

Final laboratory test results were in compliance to finished product specification, reported in the following table for the validation batches. Detailed reports attached with the batch documents.

Tests	Finished Product Specifications	S. Test Method	Test Results of Batch No.		
			001	002	003
Description	Light pink, Mottled, dumble shaped flat faced beveled edges tablets		Conform	Conform	Conform
Marking	Face one: Break line, Face Two: Break line		Conform	Conform	Conform
Dimension	(10.0 mm±0.5mm) x (5.0 mm±0.5mm)		Pass	Pass	Pass
Uniformity of weight	Target weight=140 mg/Tablet Limits: Average weight ± 5% NMT two of the individual weight deviate from the average by more than 7.5% and none should deviate by more 15%.				
Friability	NMT 1.0%		Conform	Conform	Conform
Water Content	NMT 7.0%				
Identification API	The retention times of major Peak in the chromatogram of assay preparation Correspond to that in the chromatogram of standard preparation as obtained in the assay		+Ve	+Ve	+Ve
Color (FD and C Blue) ID	Passes the test		Pass	Pass	Pass
Color (Iron Oxide) ID	Passes the test		Pass	Pass	Pass
API Dissolution 1	NLT80.0% of the LA of API dissolved in 45 minutes		%	%	%
[1]API Dissolution 2			%	%	%
API Dissolution 3			%	%	%
API Dissolution 4			%	%	%

Tests	Finished Product Specifications	S. Test Method	Test Results of Batch No.		
			001	002	003
API Dissolution 5			%	%	%
API Dissolution 6			%	%	%
Related Substances	a. API Sulfonamide: NMT 0.4%				
	b. API Urethane : NMT 0.1%				
	c. API-3-Isomer : NMT 0.2%				
	d. API unspecified individual: NMT 0.1%				
	e. API Excluding Sulfonamide: NMT 0.5%				
Content Uniformity	Each of the 10 tablets lies within the range of 85.0%–115.0% of the LA and the relative standard Deviation is NMT 6.0%		Avg =92.06% RSD=3.8%	Avg =92.418% RSD=2.68%	Avg =100.77% RSD=1.94%

1 The number of tablets could as well be 10 instead of 6 in case of validation batches. Add rows to add more results in the table.

Tests	Finished Product Specifications	S. Test Method	Test Results of Batch No.		
			013	014	015
Assay Glimepiride	1.0 mg/tablet (Limit: 90.0%–105.0% of LA)	HP-0665	0.977 mg/tab 97.7%	1.034 mg/tab 103.4%	0.990 mg/tab 99.01%
Microbial limit					
TAMC	Not more than 100 c.f.u./g	MC-0039	< 10 cfu/g	< 10 cfu/g	< 10 cfu/g
Fungi	Not more than 10 c.f.u./g	MC-0039	< 10 cfu/g	< 10 cfu/g	< 10 cfu/g
Test for pathogen	Absence of *E. coli* *Pseudomonas aeroginosa* *Staphylococcus aureus*	MC-0039	Absent	Absent	Absent

13 OBSERVATION

The manufacturing and packaging processes were reviewed during the processes running and were found within the parameters for the validation batches.

The physical parameters such as appearance, color, and consistency, and results obtained from the QC samples for these validation batches also found within specified limits.

The results obtained for the assay of active pharmaceutical ingredient of these batches also show that a quality product was produced.

14 CONCLUSION

The prospective validation study was performed on three consecutive batches of xyz 1 mg Tablets which were successfully manufactured in ABC Pharmaceutical industries.

Summarizing the results obtained during the validation for xyz 1 mg Tablets, it was concluded that:

All three validation batches for xyz 1 mg Tablets meet critical quality attributes predetermined in acceptance criteria as specified in the finished product specification.

According to the analytical results all of the criteria have been met and the process for xyz 1 mg Tablets is considered validated.

15 REVALIDATION

The production process of xyz 1 mg Tablets shall be revalidated if any critical change which can affect the quality of final product is made. The change could be but not limited to:

- Active ingredient supplier
- Mixing time
- Introducing a new equipment
- Primary packaging material

QUESTIONS AND ANSWERS

1. **What is a process validation report?**
   Answer: The process validation report is a document which demonstrates evidence that a manufacturing process is capable of consistently delivering quality products. It provides proper documentation of process validation protocols. Process validation reports are generally completed before the routine commercial production for a new product.

2. **What is the best time to publish the process validation report?**
   Answer: Process validation reports are generally completed before the routine commercial production for a new product.

3. **What are the contents of a validation report in general?**
Answer: The validation report generally contains an executive summary listing the most prominent results of the validation. A short introduction stating who produced it, when, and for what purpose. The process validation report also includes the details of the materials used with the batch numbers, the equipment with the ID numbers, and the personnel who carried out the processes with their training records ensuring that the process controls were performed by experienced operators. The raw data are further discussed in the report.

4. **Is there a need to report anomalies that happened during the validation study, in the validation report?**
Answer: If there has been any anomaly during the validation study, it is important to address it in the report with the details of investigations carried out as well as the outcome of the investigation and corrective and preventative actions taken.

8 Cleaning Validation

Cleaning validation is a procedure used to assure that cleaning processes for manufacturing equipment remove chemical and microbial residues of the active, inactive, or detergent ingredients of the product manufactured from the equipment and prevents product contamination.

Health regulatory bodies and inspectors expect pharmaceutical industries to have cleaning validation master plan to describe in detail the approach that the company will be utilizing in cleaning the manufacturing equipment between the batches of the same products and between different products. A written standard operating procedure (SOP) detailing the cleaning processes used for various pieces of equipment is also significant.

There are two basic principles in all cleaning processes:

1. **pH**
2. **Physical removal**

The best approach is physical removal that helps in removing the microbial contamination. Physical removal also lessens the numbers of pathogens from all surfaces as well as removes the food sources from the equipment surface that aide in their growth.

CLEANING VALIDATION MASTER PLAN

The Validation Master Plan is a document that includes systems, equipment, methods, facilities, etc., which are in the scope of the plan and the current validation status for the systems within the project scope. The master plan explains compliance requirements for validation, including how the validated state will be maintained. The plan also includes a matrix according to which the cleaning procedures will be validated. Once a new product or a new manufacturing equipment is introduced, the matrix is updated to include the new product/process if required.

Cleaning Program Norms

The cleaning procedures used in a facility can uncover important factors regarding process control, process reproducibility, sample collection procedures, and ways of monitoring the efficacy of cleaning procedures to first characterize the types of cleaning that are used in the facility. Cleaning methods used in the facility can tell us about the factors related to process reproducibility, process control, how to challenge the process, how to collect samples, and the best ways of monitoring cleaning efficacy during cleaning.

DOI: 10.1201/9781003366393-11

CLEANING METHODS

Automated and Manual Cleaning

There are two types of cleaning methods. Automated and manual. Manual cleaning is a normal practice in the pharmaceutical industry, yet the use of automated cleaning usually provides reproducible results. The automated system is validated by challenging the procedure and the cleaning cycle is proved to be rugged which provides reproducible results. Sometimes the design and construction of equipment make manual cleaning a necessity. Ruggedness of the method can also be given emphasis in manual cleaning; however, reproducibility depends on strict adherence to the cleaning procedures.

Clean in Place and Clean Out of Place

The Clean in Place (CIP) system involves the cleaning of large pieces of equipment at its permanent location in a configuration very similar to that utilized for production. The process is quite similar to automated cleaning, where the control system also becomes part of cleaning validation. On the other hand, smaller equipment may be transported to a designated wash area where cleaning is performed. This practice is known as clean out of place (COP).

However, using automated wash systems for COP reduces the differences between CIP and COP to a significant extent, mainly due to the reproducibility of the results.

EQUIPMENT

Dedicated and Non-Dedicated Equipment

In pharmaceutical industries, dedicated equipment is used for the production of only a single product. This practice markedly reduces the chances of cross-contamination. Where the same equipment is used for the production of a range of products, the prevention of cross-contamination between products becomes the main challenge in the cleaning validation efforts.

The cleaning of dedicated and non-dedicated equipment also gives rise to concerns. CIP systems are often used for more than one tank in a facility. Special care needs to be taken in designing CIP systems. By using appropriate valving and backflow prevention, cross-contamination can be prevented. Similarly, any circulation within the CIP system should be constructed carefully and monitored closely during routine cleaning.

Minor and Major Equipment

Major equipment represents those that play a central role in production processes. Typically, the cleaning of major equipment will be the subject of specific SOPs, and it is important to differentiate those pieces of equipment that are central to the production process from those that play a secondary role (utensils). Material of construction should be of significant importance when establishing a cleaning validation program.

Equipment should not demonstrate any type of reaction with process materials, which contact them. Equipment with porous surfaces, for example, filters, filter bags, fluid bed dryer bags, membrane filters, etc., will require thorough assessment while reviewing cleaning validation evaluations so as to ensure adequate product removal and minimize the potential for cross-contamination.

Non-Critical and Critical Site of Equipment

Locations that have a tendency to endanger a single dose with a high level of contamination are called critical sites. Such locations or sites demand special cleaning emphasis.

Besides ensuring that enough details are included in the cleaning procedure, the risk can be further reduced or completely eliminated by using more intensive sampling and testing plans. A more stringent acceptance criterion must also be established in this case to ensure effective cleaning validation.

Non-Product Contact versus Product Contact Surfaces

As a matter of course, cleaning validation mainly focuses on product contact surfaces. However, in order to be more effective, programs for the elimination of cross-contamination must also address non-product contact surfaces. When establishing the prerequisites for non-product contact surfaces, the probable interactions of that area with the process must also be reviewed.

Equipment Train: Simple and Complex

The group of equipment jointly functioning to carry out the production processes for a product is generally called "equipment train". There is a direct relationship between the complexity of cleaning validation and the complexity of the equipment train. The greater the pieces of equipment in the train, the higher the complexity of cleaning validation.

PRODUCT

Low- and High-Risk Drugs

Materials of lower pharmacological activity do not have major issues in setting residual limits for cleaning validation. However, there are materials and formulations where even minute quantities can have pharmacological activity. In such cases, although the equipment and cleaning procedures might be the same, tighter limits will be required for products with known adverse effects. Besides this, sampling and analytical methods also need to be refined to a high degree of sensitivity to ensure the removal of residue from equipment.

Solid and Liquid Dosage

Liquid product has a greater ability to penetrate equipment seals and joints, whereas solid product can form tufts or clumps on the surface, which may prevent wetting of that part by cleansing agents and thus inhibit the ability to rinse the residues properly. The same phenomenon can be true for the dispersion of contaminants on the surface of the equipment for solid and liquid products. The distribution of contaminants in the case of solid products may vary from point to point while that in liquid products is uniform across the surfaces.

Soluble and Insoluble Ingredients

The removal of insoluble materials from the equipment surface represents yet another difficult scenario because it requires more physical means as compared to soluble materials (active or excipient), which are often easily removed by solubilizing the product.

Sterile and Non-Sterile Facilities

Sterile manufacturing facilities differ from non-sterile ones because of the extra precautions required to control microbial and endotoxin levels. In non-sterile products, environmental concerns are reduced but are still important.

FACILITY

Single-Product Facility and Multiple-Product Facility

Since no cross-contamination concerns exist in the case of a facility producing only a single product, the validation requirements are also minimized. However, careful monitoring of changeover of equipment from one product to another has to be done.

In addition, continuous monitoring must also be warranted to ensure that all controls and limits established are in place after accomplishment of cleaning validation.

Campaign Production and Batch Production

In a multiple-product facility, campaign lots of a single product or product family are produced in the same equipment which minimizes chances of cross-contamination. At times, the production trot may be stopped for a part cleanup of the equipment, which is less stringent than a full cleanup. Once the campaign production is over, an intensive cleaning of the facility and equipment can be performed before starting the production of a different product.

CLEANING VALIDATION

A master plan is the basis of the cleaning validation program, which describes the overall approach of cleaning validation. This includes the matrixing philosophy involved and the rationale associated thereto. Once the products and pieces of equipment are identified for use in the validation study, trials may start.

Some worst-case scenarios may also be considered to challenge the cleaning procedure, for example, having the product dried on the surface to make the cleaning difficult or applying the effect of weekends and holidays on the cleaning schedule, and so on.

A brief review of the activities to establish a comprehensive cleaning validation program is given below.

CLEANING VALIDATION PROGRAM

a. *Product grouping:* Based on formulation and dosage form, potency, toxicity, and solubility, all products are grouped. The broad groups may then be divided into subgroups according to formulation and process types. After the grouping, the worst-case product is selected from each group. The worst-case product from each group may be the least soluble, the most toxic, or with the highest concentration of active ingredients. In some situation, a combination of these parameters may also be used.
b. *Equipment grouping:* Equipment of similar design and function is typically collected in one group for validation study. In case of similar cleaning procedures implemented, validation can be conducted on the largest- and smallest-scale equipment separately.
c. *Cleaning methods grouping:* The grouping of cleaning procedures may be appropriate; however, the validation of the cleaning procedure may also be conducted independently of the equipment for which it is used.
d. *Cleaning agents grouping:* Systems may also be subdivided on the basis of cleaning agents utilized on those systems when considering product formulation and equipment groupings. Incidentally, the use of a single cleaning agent will greatly minimize the work required to determine if residues of the agent remain after cleaning.

RESIDUES AND RESIDUE REMOVAL

a. *Types of residues:* Physical and chemical properties such as solubility, hydrophobicity, and reactivity of residues affect the ease with which they are removed from surfaces. It is therefore important to first identify the substance to be cleaned.
b. *Cleaning agents:* It is necessary to know the ingredients of a cleaning agent because when cleaning agents are used to aid cleaning, their removal must also be demonstrated to ensure the proper cleaning of surfaces. Once the ingredients are known, validation personnel must then determine the worst-case ingredient in the cleaning agent.

CLEANING OF EQUIPMENT

a. *Types of cleaning processes (manual/semiautomated/automated):* The direct cleaning of equipment is considered manual cleaning. Among the critical parameters involved in manual cleaning are volumes of cleaning agents and rinse water, temperature of wash and rinse solutions, duration of washing cycle, concentration of detergent used, and so on.

In semiautomated cleaning, various levels of automatic control are also included. This may also be considered as a blend of manual and automated cleaning, for example, manually removing gaskets and fittings before automated CIP or dismantling a pump prior to cleaning in an automated COP system. The automated system usually comprises programmable cycles and does not include personnel intervention.

b. *CIP/COP:* As discussed in the sections above, CIP generally refers to the automated circulation system.
c. *Equipment design considerations:* Care must be taken in designing equipment to minimize the risks of cross-contamination and microbial contributions to the equipment. Preferably, equipment should be constructed of a non-reactive material.
d. *Equipment storage after cleaning:* It is necessary to protect equipment from cross-contamination between the period of cleaning completion and reuse for the next product manufacture. Areas must be allocated for this purpose where possible cross-contamination may be controlled. In the meantime, records must also be maintained showing equipment numbers, date and time of cleaning, and names of persons who cleaned and inspected it.

CYCLE DEVELOPMENT

a. *Cleaning agent selection:* Selection criteria for cleaning agents should be the suitability of removing product residues and low toxicity. Besides these, ingredients of the selected cleaning agent should also be known so that the cleaning of reagent itself can be proven.
b. *Cleaning parameter selection:* The most important cleaning parameters are time, temperature, cleaning agent concentration, and cleaning action, for example, impingement, sheeting, rinsing, and so on. By evaluating each cleaning step, the removal of residues can be determined and thus the need to add, delete, or modify a cleaning step can be decided as well.
c. *Standard operating procedures:* A draft-cleaning procedure should be in place prior to starting the cleaning validation. Once a successful validation is accomplished, the final SOP for cleaning must be completed with details such as time, temperature, concentration, and cleaning action.
d. *Operator training:* A formal training of operators includes reviewing and understanding the cleaning SOPs, qualified apprenticeship, and ensuring that training is successful. Operators must also understand the process of cleaning and the equipment they are cleaning.

SAMPLING TECHNIQUES AND ANALYTICAL METHODS

a. *Swabs and wipes:* Swabs and wipes are widely accepted sampling techniques. Their advantages are that they dissolve and physically remove samples, are economical, allow sampling of the defined area, are usable on a variety of surfaces, and are applicable to active ingredients, microbial, and cleaning agents. However, there are some limitations involved with swabs and wipes: for example, they may introduce fibers and material to the sampling area; sometimes the design of the swab may also inhibit the recovery and specificity of the method; and they are difficult to use in crevices, pipes, or large vessels.

b. *Rinse sampling:* The advantages of rinse sampling are the following: ease in sampling, coverage of large areas in samples including sampling of unique surfaces, being adaptable to on-line monitoring and fewer technicalities involved than swabs, and so on. Restrictions include a possible decrease in test sensitivity, inability to detect residue locations, inadequate homogenization of residues, and minimum information about actual surface cleanliness in some cases. Due to the criticality of rinse volume, usually the entire piece equipment is used for rinsing, such as a vessel.
c. *Direct surface monitoring:* The benefits of direct surface monitoring are that it is fast, non-invasive, and economical. There are some limitations however; for example, there are some prejudices and some techniques which are not available yet. Visual examination of equipment for cleanliness immediately before use is a requirement by CGMP regulations. It is a form of direct surface analysis. Other commonly used methods of monitoring include pH, conductivity, total organic carbon (TOC) titration, high-performance liquid chromatography (HPLC), thin-layer chromatography (TLC), capillary zone electrophoresis, Fourier transform infrared (FTIR), atomic absorption, ultraviolet (UV) spectrophotometry, and so on.

LIMITS AND ACCEPTANCE CRITERIA

The most important element of a good cleaning validation program is the determination of limits and acceptance criteria. When determining the limits, care must be taken so that they are achievable by the analytical methods available for the specific product and active ingredient, are practical for the actual cleaning situation to be validated, and are scientifically rationalized and verifiable.

The most commonly used basis for setting the acceptance limit is a mathematical calculation that allows a certain therapeutic dose to carry over into each dosage unit of the next product. The actual numerical limits are based on the pharmacological potency of the product, the toxicity of the residue, and the analytical limit of detection.

ONGOING MONITORING OF CLEANING

Besides inspection of each piece of equipment to ensure cleanliness before use, additional verification can also be done. This depends largely on the complexity of the equipment.

Automated cleaning methods may not require ongoing verifications; however, semiautomated processes and manual cleaning usually need periodic verification and determination about the reproducibility of the process over time.

CHANGE CONTROL

Changes made to cleaning SOPs, analytical methods, detergents, equipment, product formulation, etc. should fall under the auspices of the change control policy of the company.

Formal documentation will be required to make changes to these items. Changes performed under the change control policy will require reconfirmation of the original cleaning validation results. In case the change is deemed to be fundamental to the grouping philosophy or to the cleaning method, the change may require a revalidation, which may differ from verification only by the amount of sampling.

QUESTIONS AND ANSWERS

1. **What is the major concern of using non-dedicated equipment in cleaning Validation?**
   Answer: When the same equipment is used for the production of a range of products, the prevention of cross-contamination between products becomes the main challenge in the cleaning validation efforts. Dedicated equipment is used for the production of only a single product. This practice markedly reduces the chances of cross-contamination.

2. **Describe the term "Product grouping" in cleaning validation and how the worst case is selected?**
   Answer: Based on formulation and dosage form, potency, toxicity, and solubility, all products are grouped. The broad groups may then be divided into subgroups according to formulation and process types. After the grouping, the worst-case product is selected from each group. The worst-case product from each group may be the least soluble, the most toxic, or with the highest concentration of active ingredients. In some situation, a combination of these parameters may also be used.

3. **Explain why storage after cleaning of equipment is critical and what measures need to be taken to avoid cross-contamination of a cleaned equipment?**
   Answer: It is necessary to protect equipment from cross-contamination between the period of cleaning completion and reuse for the next product manufacture. Areas must be allocated for this purpose where possible cross-contamination may be controlled. In the meantime, records must also be maintained showing equipment numbers, date and time of cleaning, and names of persons who cleaned and inspected it.

4. **What are the types of sampling in cleaning validation?**
   Answer: There are three types of sampling in cleaning validation;
   1. *Swabs and wipes:* Swabs and wipes are widely accepted sampling techniques. Their advantages are that they dissolve and physically remove samples, are economical, allow sampling of the defined area, are usable

on a variety of surfaces, and are applicable to active ingredients, microbial, and cleaning agents.

2. *Rinse sampling:* The advantages of rinse sampling are ease in sampling, coverage of large areas including sampling of unique surfaces, being adaptable to on-line monitoring and fewer technicalities involved than swabs, and so on. Due to the criticality of rinse volume, usually the entire piece equipment is used for rinsing, such as a vessel.
3. *Direct surface monitoring:* The benefits of direct surface monitoring are that it is fast, non-invasive, and economical.

5. How do we differ in CIP and COP systems?

The CIP system involves the cleaning of large pieces of equipment at its permanent location in a configuration very similar to that utilized for production.

On the other hand, smaller equipment may be transported to a designated wash area where cleaning is performed. This practice is known as clean out of place (COP).

However, using automated wash systems for COP reduces the differences between CIP and COP to a significant extent, mainly due to reproducibility of the results.

9 Method Development and Method Validation

METHOD DEVELOPMENT

Method development is the process of proving that an analytical method is acceptable for use to measure the concentration of an API in a specific compounded dosage form which allows simplified procedures to be employed to verify that an analysis procedure, accurately and consistently, will deliver an accurate measurement of an active ingredient in a dosage form or raw material.

The analytical method validation is essential for development and has been tested extensively for specificity, linearity, accuracy, precision, range, detection limit, quantitation limit, and robustness. In summary, analytical method development and validation allow confirming that an accurate and reliable potency measurement of a pharmaceutical preparation can be performed.

The number of drugs introduced into the market is increasing day by day. These drugs may be either new or derivative of the existing ones.

ANALYTICAL METHOD DEVELOPMENT

The most common methods in the pharmaceutical industry are HPLC or UPLC methods.

The basic separation techniques and principles involved in the analytical method development using the HPLC and UPLC are listed as follows:

- Selection of chromatography mode
- Selection of detector
- Selection of column (stationary phase)
- Selection of mobile phase, including a selection of buffer, pH of buffer, and composition of the mobile phase.
- Selection of ion-pair reagents
- Optimization of flow rate
- Selection of diluent
- Standard and sample preparation
- Selection of injection volume
- Finalization of wavelengths
- Stability of solution
- System suitability

Analytical method validation deals with methods for identifying, separating components, and quantifying natural and artificial materials. The choice of analytical

DOI: 10.1201/9781003366393-12

methodology is based on many considerations, such as the chemical properties of the substance and its concentration type of measurements, i.e., quantitative or qualitative and the number of samples. A qualitative method yields information on the chemical identity of the species in the sample. A quantitative method provides numerical data regarding the relative amounts of one or more of the analytes in the sample.

High-Performance Liquid Chromatography (HPLC)

The HPLC unit contains the following parts:

- **Column**

  The column is the stationary phase. To select the column, find out the column used for similar compounds. The dimension can be modified accordingly. If the compound is polar, use the column with non-polar material, and if the compound is non-polar, use the column with polar material and check the peak shape; if the peak is broad, use a smaller column of the same material. If the peak is in good shape but coming late and cannot control with other parameters, use a longer column.

  The columns are made of silica, which is polar in nature. A column with polar material is also known as a normal phase column.

  The silica for those materials is modified, introducing long-chain carbons like C-8 or C-18. Those columns are non-polar in nature and known as reverse-phase columns.

  Sometimes several columns have to be tried to get a good chromatogram.
- **Detector**

  The most common detectors are UV detectors. Most organic compounds are detected using UV detectors. Other detectors are Refractive index (RI) detectors, Florescence detectors, etc. Those are used when compounds are not detectable by UV.

  You need a wavelength to detect any compounds. The solution of a compound can be scanned with a UV spectrophotometer to determine its wavelength.

  For the greatest sensitivity, λ_{max} should be used to detect all sample components containing chromophores. UV wavelengths below 200 nm should be avoided because detector noise increases in this region. Higher wavelengths give greater selectivity.
- **Pump**

  The pump flows towards the mobile phase through the column and detector.

  Usually, the flow rate is 0.5–5.0 mL/min. The most common flow rate is 1.0–2.0 mL/min. If the peak is coming late and not in good shape, the flow rate can be increased to bring the peak earlier and to a better shape. The column backup pressure should not exceed the limit of the system. If backpressure is too high, the flow rate should be lowered.

The flow rate should be at most 2.5 mL/min. Select the flow rate with the least retention time, good symmetrical peak shape, lower back pressure, and better separation between two components.

- **Injector**

  The injector loads the sample into the column. The most common volumes are 10–200 µL. If the sample is concentrated, use a smaller volume of injection. If the peak is too small, use a larger injection volume.

- **Column heater**

  Column heaters are close components in which columns are placed during chromatography. By using heaters with higher temperatures, peak shape can be improved. Higher temperatures bring the peak earlier, and a sharper peak is produced.

- **Selection of mobile phase**

  Again, check the similar compounds to find out the correct mobile phase.

  The solvent strength is a measure of its ability to pull analytes from the column. It is generally controlled by the concentration of the solvent with the highest strength; for example, in reverse phase HPLC with aqueous mobile phases, the strong solvent would be the organic modifier; in normal phase HPLC, it would be the most polar one. The aim is to find the correct concentration of the strong solvent. With many samples, there will be a range of solvent strengths that can be used within the aforementioned capacity limits. Other factors (such as pH and the presence of ion pairing reagents) may also affect the overall retention of analytes.

  For polar compounds, a polar mobile phase is used, and for non-polar compounds, non-polar mobile phases are used.

  Mobile phases are a mixture of organic solvents (usually methanol and acetonitrile) and buffers.

  Mobile phases with low pH are used for basic compounds, and for acidic compounds, a mobile phase with higher pH is used.

  Some ion pair reagents are also used in the mobile phase. Sodium salts of alkyl halides produce a negative charge in the mobile phase to attract the positively charged components. Amines are added to the mobile phase to produce a positive charge in the mobile phase.

  A buffer is a partially neutralized acid which resists changes in pH. Salts such as Sodium Citrate or Sodium Lactate are normally used to neutralize the acid partially. Buffering Capacity is the ability of the buffer to resist changes in pH.

  Buffering Capacity increases as the molar concentration (molarity) of the buffer salt/acid solution increases.

  Consideration of the effect of pH on analyte retention, the type of buffer to use, its concentration, solubility in the organic solvents, and its effect on detection are essential in the reversed-phase chromatography (RPC) method development of ionic analytes. An improper choice of buffer, in terms of buffering species, ionic strength and pH, can result in poor or irreproducible retention and tailing in the reverse-phase separation of polar and ionizable compounds.

Phosphate is more soluble in methanol/water than in acetonitrile/water or THF/water. Ammonium salts are generally more soluble in organic/water mobile phases.·TFA can degrade with time, is volatile, and absorbs at low UV wavelengths.

After buffers are prepared, they should be filtered. Mobile phases should be degassed.

Generally, a 10–50 mM buffer concentration is adequate for small molecules. Generally, up to 50% organic should be used with a buffer. This will depend on the specific buffer as well as its concentration. Phosphoric acid and sodium or potassium salts are reverse phase most common buffer systems. Phosphonate buffers can be replaced with sulfonate buffers when analyzing organophosphate compounds.

Selection of the Diluting Solvent (Diluent)

The diluent should be capable of dissolving both the standard and the sample.

Organic solvents or a mixture of organic solvents can be used as diluents. If compounds are insoluble in the organic mixture, the buffer can be added depending on the nature of the compounds.

The drug substance being analyzed should be stable in solution (diluent). During initial method development, the solutions in amber flasks should be prepared until it is determined that the active component is stable at room temperature and does not degrade under standard laboratory conditions. The sample solution should be filtered; using a 0.22 or 0.45 μm pore-size filter is generally recommended for removing particulates. Filtration is a preventive maintenance tool for HPLC analyses.

Sample preparation is a critical step of method development that the analyst must investigate. The effectiveness of the syringe filters is largely determined by their ability to remove contaminants/insoluble components without leaching undesirable artifacts (i.e., extractable) into the filtrate. If any additional peaks are observed in the filtered samples, the diluent must be filtered to determine if a leachable component comes from the syringe filter housing/filter.

Sample Preparation

The prepared sample should be an aliquot relatively free of interferences that is compatible with the HPLC method and will not damage the column. The main sample preparation techniques are liquid-liquid extraction and solid-phase extraction. In these methods, the analyte of interest was separated from the sample matrix so that as few potentially interfering species as possible were carried through to the analytical separation stage.

System Suitability

Whether a chromatographic system is being routinely utilized today in pharmaceutical laboratories where the quality of results is most important is suitable for a definite analysis. Those limits for system suitability ensure that a system will produce consistent results.

The parameters used in the system suitability test (SST) report are as follows:

1. Number of theoretical plates or Efficiency (N).
2. Capacity factor (K).
3. Separation or Relative retention (α).
4. Resolution (Rs).
5. Tailing factor (T).
6. Relative Standard Deviation (RSD).

Those parameters are calculated by using the software.

Ultra-Performance Liquid Chromatography (UPLC)

Ultra-performance liquid chromatography (UPLC) is a recent technique in liquid chromatography that enables significant reductions in separation time, solvent consumption, and analysis time compared to conventional HPLC.

METHOD VALIDATION

After developing an analytical procedure, it is most essential to ensure that the procedure consistently produces the intended precise result with high accuracy. The method should give a specific result that may not be affected by external matters. This creates a requirement to validate the analytical procedures. The validation procedures consist of some characteristics and parameters that make the method acceptable with the addition of statistical tools.

The following factors are typical analytical performance characteristics which may be tested during methods validation:

1. Specificity
2. Accuracy
3. Precision
4. Linearity
5. Detection of limit
6. Quantitation limit
7. Robustness
8. Force Degradation
9. Stability of samples and standards

Specificity

Developing a separation method for HPLC involves demonstrating specificity, which is the ability of the method to accurately measure the analyte response in the presence of all potential sample components. The response of the analyte in test mixtures containing the analyte and all potential sample components (placebo formulation, synthesis intermediates, excipients, degradation products, and process impurities) is compared with the solution's response containing only the analyte.

In this step, the mobile phase, diluent, and placebo are injected using the same parameter as a sample. There should be no peak from those which can interfere with the active peak.

Accuracy

Accuracy is the nearness of a measured value to the actual or accepted value.

The accuracy of an analytical procedure expresses the closeness of agreement between the value, which is accepted either as a conventional true value or an accepted reference value and the value found.

Precision

The precision of an analytical procedure expresses the closeness of agreement between a series of measurements obtained from multiple samplings of the same homogeneous sample under the prescribed conditions. Precision should be carried out using homogeneous samples. The precision of an analytical procedure is usually expressed as the variance, standard deviation, or coefficient of variation of a series of measurements.

In this experiment, six or more samples are prepared and tested using that method. All values must be close to the actual value and close to each other. Passing criteria can be established i.e., 2% RSD for the six replicate results.

Intermediate precision results from variations such as different days, analysts, and equipment. Experimental design should be employed to determine intermediate precision so that the individual variables' effects (if any) can be monitored. Precision criteria for an assay method are that the instrument and intra-assay precision (RSD) may be $\leq 2\%$.

Repeatability expresses the precision under the same operating conditions over a short time interval. Repeatability is also termed intra-assay precision.

Reproducibility expresses the precision between laboratories (collaborative studies, usually applied to standardization of methodology).

Linearity

The linearity of an analytical procedure is its ability (within a given range) to obtain test results directly proportional to the concentration (amount) of the analyte in the sample.

If a target for the assay sample is 100%, a 50%–150% series may be prepared and tested (ICH requirement is 80%–120%).

The acceptability of linearity data is often judged by examining the correlation coefficient and y-intercept of the linear regression line for the response versus concentration plot. The regression coefficient (r) is.0.998 and is generally considered evidence of an acceptable fit of the data

DETECTION OF LIMIT

The detection limit of an individual analytical procedure is the lowest amount of analyte in a sample which can be detected but not necessarily quantitated as an exact value.

In this procedure, the area of noise is calculated and the Limit of Detection is considered three times the area as compared to noise.

QUANTITATION LIMIT

The quantitation limit of an individual analytical procedure is the lowest amount of analyte in a sample which can be quantitatively determined with suitable precision and accuracy. The quantitation limit is a parameter of quantitative assays for low levels of compounds in sample matrices and is used mainly to determine impurities and/or degradation products.

A series of low-concentration samples are prepared and tested; the curve is drawn. The lowest point of the curve is known as the "Limit of Quantification". Then the LOQ solution is tested for precision.

Robustness

The robustness of an analytical procedure is a measure of its capacity to remain unaffected by minor, deliberate variations in method parameters and indicates its reliability during normal usage.

For example a minor change is made and samples are tested; the results should not be affected:

The flow rate of HPLC may be altered by 10% lower and higher.
The mobile phase combination may be altered. The combination may change if the mobile phase is a mixture of organic and aqueous.
The wavelength may be altered to ± 3 nm.
The column temperature may be altered to ± 2°C.

Force Degradation

Samples are exposed to drastic conditions, such as exposure to high temperatures and exposure to light, and treated with an oxidizing agent like hydrogen peroxide. The method should be capable of testing the exposed sample.

Stability of Samples and Standards

The two sets of samples should be prepared. One should be kept at room temperature and another in the refrigerator.

The sample should be tested as soon as prepared; then each set the samples (room temperature and refrigerator) should be tested after one day, two days, and three days. The stability of the samples/Standard can be determined by comparing the results from the initial test.

QUESTIONS AND ANSWERS

1. **What are the common components of an HPLC system?**
   Answer: Column, detector, pump, injector, and column heater.

2. **What will be the effect of high column temperature on peak retention time?**
   Answer: The peak will elute earlier.

3. **If a compound is acidic in nature, and the peak is eluting late, what is to be changed in pH to elute its peak earlier?**
   Answer: High pH (basic) will pull peak earlier.

4. **What are the basic parameters for analytical method validation?**
   Answer: Specificity, Accuracy, Precision, Linearity, Limit of Detection, Quantitation limit, Robustness, Force Degradation, Stability of samples, and standards.

5. **What is the Limit of Quantification?**
   Answer: The lowest amount, quantified by a method, is known as the limit of quantification.

6. **What is Robustness?**
   Answer: The robustness of an analytical procedure is a measure of its capacity to remain unaffected by minor, deliberate variations in method parameters and indicates its reliability during normal usage.

10 Qualification and Its Segments

Qualification can be defined as the process of establishing documented evidence to ensure that a specific equipment, facility, or system is fit and ready for their intended use. Equipment qualification can also be defined as the activities in which various inspections, tests, and assessments of the entire system are performed to ensure that a given piece of equipment is compliant and ensures reliable performance. It further confirms that the equipment will continue to do, on a consistent basis, what it is supposed to do.

To elaborate further, equipment qualification is the documented confirmation of the capability of the system to consistently operate within set limits and tolerances, and allows for conduction of tests to be performed on one element or component of the process to be validated against a specified outcome.

Validation, verification, and qualification are the terms which are used more frequently in the pharma industry. Let us first understand the difference between the three concepts.

See below:

Validation: Validation is a process or instance to support a system or process on a sound authoritative basis that the specified requirements for a specific intended use or application have been fulfilled.

Verification: Verification is the process of establishing the reality of a system through the provision of objective evidence that the specified requirements have been fulfilled. For example, comparison of a new design specification with a similar proven design specification.

Qualification: Qualification is a process to assure that a system under question complies with some condition or pre-established standards and/or specific requirements. Qualification can concern person, products, processes, or systems.

Since in this chapter, we are focusing more on qualification (or equipment qualification), let us magnify our information about the four equipment qualification stages. The four qualification stages help determine if the process being tested is capable of operating in its specific environment.

These stages are:

- Design Qualification (DQ)
- Installation Qualification (IQ)

DOI: 10.1201/9781003366393-13

- Operational Qualification (OQ)
- Performance Qualification (PQ)

The following are the details of the four stages of qualification.

DESIGN QUALIFICATION

Design qualification ensures to provide documented verification that the proposed design of a new equipment is suitable for the intended purpose. It is a common part of equipment qualification and is the process of documenting design reviews to illustrate that all quality aspects that involve the activities of purchasing the equipment and supplies work have been considered at the design stage.

WHY TO DO DQ

The question why to perform a design qualification is easy to answer. First of all, DQ enhances the reliability of the equipment. It is a tool that can help the user achieve compliance in the equipment qualification process. By challenging the design of the equipment through the design qualification process the user can confirm that the design of the equipment is fit for its intended purpose and function, meets the expectations, and complies with relevant standards, guidelines, and regulations. It can further reduce the risk of errors, failures, or defects in the equipment while decreasing the cost and time required for installation, operational, and performance qualification studies of the system.

DQ PROTOCOL

A DQ protocol needs to be prepared and approved before the design of the equipment is finalized, and should be based on a risk assessment and a design review.

In the DQ protocol, the scope of the document is defined as to what this DQ overall encompasses. Besides scope, objectives, criteria, methods, and responsibilities are other areas to be described in the protocol for conducting the DQ. The DQ protocol also provides what the acceptance criteria would be to conduct the tests against, the test procedures, and verification activities, to demonstrate that the design of the equipment meets the user requirements and specifications.

DQ REPORT

Once the protocol has been executed, the data and results are collected and are compiled in a report to draw conclusion. The DQ report should be prepared and approved, containing elements such as a summary of the objectives, scope, criteria, methods, and responsibilities with reference to the approved protocol.

If during or before execution of the protocol, any amendments to the protocol were done that will also be mentioned in the report. A brief description of the equipment design, list of user requirements specifications, summary of test procedures, verification activities and acceptance criteria, test results, data, and observations will be essential segments in the DQ report. This report provides evidence that the design of the equipment meets the user requirements and specifications set in the DQ protocol. In case of any deviations or non-conformances during the DQ study, the report needs to identify them. It should also explain as to what corrective measures, if any, were taken to mitigate the deviations.

INSTALLATION QUALIFICATION

Installation qualification is actually a process to demonstrate that the equipment being installed will offer a high degree of quality assurance such that production processes run on that system will consistently produce a product that will meet all pre-established quality requirements.

Installation Qualification (IQ) also verifies that an instrument or unit of equipment and its associated sub-systems has been installed and configured according to the pre-established specifications and installation checklist. This may apply to both new equipment or after a repair or change of any of the parts of the system.

In addition to process validation, equipment validation (or qualification) is also very for the manufacture of a high-quality product consistently. The key sets of protocols within equipment validation are Installation Qualification (IQ), Operational Qualification (OQ), and Performance Qualification (PQ). The other two components will be discussed further in this chapter after IQ.

Let's first understand the concept of installation qualification (IQ).

As an example, consider the scenario of a physical instrument that requires a specific amount of floor space, certain operating conditions, and an assurance that no damage exists on the unit. To make the IQ successful, one needs to ensure that the following conditions are met as per specifications and the approved qualification protocol:

- Location of the unit installed and necessary floor space and conditions
- Environmental and operating conditions
- Availability of all operating and safety manuals and certifications (where applicable)
- List of all spare parts and other consumables
- List of approved lubricants and oil (where applicable)
- Documentation of computer-controlled instrumentation (if applicable)
- Documenting firmware and software versions and serial numbers
- De-boxing of the unit and cross-checking instruments and components for damage
- Availability of correct power supply and connection
- Verifying connection and communication with peripheral units
- Installing ancillary instruments
- Checking software system installation and accessibility
- Checking all associated instruments for calibration certificates and inclusion in the company's site calibration program
- Recording calibration and validation dates of tools used for IQ

There are certain rules that should be followed while installing a new equipment.

Failure to follow these rules may result in serious personal injury or damage to machine.

When using the machine, basic safety precautions should be followed to reduce the risk of fire, electric shock, and personal injury including the following:

- KNOW YOUR MACHINE: The machine/system must be operated by person(s) familiar with the respective process and technology.
- LIFT MACHINE using a suitable lifting device. Always refer to the layout/assembly drawing for the approximate weight of the machine.
- DO NOT USE IN A HAZARDOUS ENVIRONMENT. Know the rating of the area where the machine is to be used and consult the specifications and layout drawing for rating of electrical components and location of sub-systems.
- GUARD AGAINST ELECTRIC SHOCK. All electrical connections must be made by a qualified electrician and must conform to local electrical codes and grounding practices. Before connecting the machine to a power source, be sure that the voltage supply is the same as specified on electrical components.
- SECURE REMOVABLE PARTS prior to starting the machine.
- NEVER STOP THE MACHINE WITH THE INTERLOCK SWITCHES use the stop button, wait for rotating parts (if any) to come to a complete stop.
- INSPECT INTERLOCK SWITCHES and cables regularly. Replace if damaged or defective. Ensure that interlock switches are working correctly prior to using the machine.
- DISCONNECT MACHINE and lock out power before opening, adjusting, cleaning.

  Servicing, and changing parts.
- STAY ALERT when using the machine.

OPERATIONAL QUALIFICATION

The process of equipment qualification that follows after installation and installation qualification is called "operational qualification". Operational Qualification (OQ) can be defined as the process which involves identifying and inspecting equipment features that can impact final product quality. OQ is actually a series of tests which confirms that equipment and its associated instrument and sub-systems will operate within their specified limits consistently and dependably.

To perform an operational qualification, just like in the case of installation qualification, an approved protocol is required that defines all the collection of test cases that will be used to verify the proper functioning of a system. The operational qualification protocol has to be approved before execution where all parties agree on the test requirements given in the protocol. The test requirements are already defined in another document which is called the Functional Requirements Specification. The functional requirement specification is written and approved before the purchase of the equipment. Operational qualification is usually performed before the system is released for use or for performance qualification (the third and final significant step of equipment qualification process).

Depending on the user's needs and the complexity of the system, operation qualification can be combined with Installation Qualification or Performance Qualification. For example, if the qualification is performed for a small system like a balance, pH meter, or water batch, the three stages of qualification studies could be merged together in one or two documents. However, for major systems like a furnace, incubator, manufacturing vessel, a tablet press, all three segments need to be performed separately.

Operational Qualifications should be approved before protocol execution. A copy of the unexecuted protocol should be kept in the validation package. The unexecuted protocol should be approved by the System Owner and Quality Assurance as a minimum. The executed protocol should be signed by the tester and reviewed by the system owner and Quality.

PERFORMANCE QUALIFICATION

Performance Qualification or PQ is performed after the completion of Operational Qualification. PQ study provides documented verification that equipment or a system is capable of performing the activities of a process according to pre-approved specifications while operating in its specified operating environment.

Depending on situation and types and complexity of the system, performance qualification can often be combined with Installation Qualification or Operational Qualification.

Similarly to IQ and OQ, a PQ should also be approved before protocol execution. The unexecuted protocol should be approved by the system owner, engineering department, and quality assurance. The executed protocol should be signed by the tester and reviewed by the system owner with quality as the final approver.

Each step of the qualification should include an instruction to conduct the specific test, an expected result, and the actual result that must be lying with in the specified range of expected results. Any discrepancy between the expected result and the actual result should be tracked as a deviation. Deviations should be resolved with involvement of quality before validation is complete.

A performance qualification will help to demonstrate:

- That a system can handle multiple users without a significant system lag
- That when the system contains large quantities of data, queries are returned in a certain (short) period of time
- That concurrent independent work-flows do not affect each other
- That a laboratory test correctly identifies a known material
- That a process was completed within defined system requirements

QUESTIONS AND ANSWERS

1. **Define qualification**
   Answer: Qualification can be defined as the process of establishing documented evidence to ensure that a specific equipment, facility or system are fit and ready for their intended use. Equipment qualification can also be defined as the activities in which various inspections, tests, and assessments of the entire system are performed to ensure that a given piece of equipment is compliant and ensures reliable performance. It further confirms that the equipment will continue to do, on a consistent basis, what it is supposed to do.

2. **How many stages of qualification are there?**
   Answer: There are four stages of qualification that help determine if the process being tested is capable of operating in its specific environment. These stages are:

   - Design Qualification (DQ)
   - Installation Qualification (IQ)
   - Operational Qualification (OQ)
   - Performance Qualification (PQ)

3. **What is the difference in validation, verification and qualification?**
   Answer: Validation is a process or instance to support a system or process on a sound authoritative basis that the specified requirements for a specific intended use or application have been fulfilled. Whereas verification is the process of establishing the reality of a system through the provision of objective evidence that the specified requirements have been fulfilled. For example, comparison of a new design specification with a similar proven design specification. Qualification differs from validation and verification in terms of the nature of the process that is executed to assure that a system under question complies with specific requirements. Qualification can concern persons, products, processes or systems.

4. **What is the definition of performance qualification?**
   Answer: Performance qualification is defined by FDA as "Establishing confidence through appropriate testing that the finished product or process produced by a specified process meets all release requirements for functionality and safety and that procedures are effective and reproducible".

   Performance qualification is the executed test protocol documenting that a system meets the defined requirements to function in the production environment.

5. **What is a qualification protocol intended for?**
   Answer: An approved qualification protocol is required to define the test cases that will be used to verify the proper functioning of a system. The qualification protocol has to be approved before execution where all parties agree on the test requirements given in the protocol

6. Describe briefly the general rules that should be looked into before installation of an equipment?

When installing a new equipment, basic safety precautions should be followed to reduce the risk of fire, electric shock and personal injury including the following:

1. The machine/system must be operated by person(s) familiar with the respective process and technology.
2. Always refer to the layout/assembly drawing for the approximate weight of the machine.
3. Know the rating of the area where the machine is to be used and consult the specifications and layout drawing for rating of electrical components and location of sub-systems.
4. All electrical connections must be made by a qualified electrician and must conform to local electrical codes and grounding practices.
5. Before connecting the machine to a power source, be sure that the voltage supply is the same as specified on electrical components.
6. Use the stop button to stop the machine and wait for rotating parts (if any) to come to a complete stop.
7. Ensure that interlock switches are working correctly prior to using the machine.
8. Lock out power before opening, adjusting, cleaning, servicing and changing parts.

7. Initiate an installation qualification protocol for a Production equipment.

Answer: The installation qualification protocol is generated as follows:

Title	______________________
Purpose	______________________
Equipment	______________________
Process/Product	______________________
Component	______________________
Location	______________________
Department	______________________
Documents Included	______________________

Completion of the following signature block signifies that the appropriate parties have reviewed this installation qualification report and agree that it is correct and complete.

Protocol Author: ______________________ Date: ___________

Reviewer and Approvers NAME/TITLE	Signature	Date
Manager Validation		
Manager Engineering		
Manager Quality Assurance		
Director Quality Operations		

TABLE OF CONTENTS

Title Page
Table of Contents

1. Introduction
   1.1 Background
   1.2 Purpose
   1.3 Scope
   1.4 References
   1.5 Exceptional Conditions and Deviations
   1.6 Validation Approach

2. Evaluations
   2.1 General
   2.2 Equipment and System Utilities
   2.3 Operation and Maintenance Program
   2.4 Instrument Calibration Program
   2.5 Lubricants
   2.6 Engineering Drawings
   2.7 Controls, Alarms, and Interlocks
   2.8 Initial Cleaning

Attachments
Attachment 1
Attachment 2

1 INTRODUCTION

1.1 Purpose

Installation qualification identifies an item or system, how it was actually installed and what SOPs will be used to operate and maintain it in the operations environment

1.2 Background

The reader needs to know what is being qualified, how it will be used, why it was selected for this service and where it will be used. One or two paragraphs should suffice.

This description may include (but is not limited to):

- *Equipment name and model*
- *Basic capacity and ratings*
- *Fixed or portable operation*
- *New or used equipment (with previous use)*
- *Planned use*
- *Why this item was selected for this service and why we feel that it will meet our needs*
- *Why the item requires validation*

This section should include a basic description of the equipment and its intended use. Some examples are as follows:

- *Electric pump to transfer water in manufacturing.*
- *Jacketed stainless steel pressure vessel for boiling water in the process.*
- *Clean steam generator to convert Purified Water, USP into clean steam using plant steam as the heat source. Attached to fluid bed dryers.*

1.3 Scope
1.4 References
1.5 Exceptional Conditions and Deviations
1.6 Validation Approach

2 EVALUATIONS

2.1 General

Complete data sheets in ATTACHMENT I including:
Equipment and System Description
Manufacturers Specifications
Purchase Orders
Materials in Product Contact

2.2 Equipment and System Utilities

Complete data sheets in ATTACHMENT I corresponding with the applicable items below.

	Applicable	Not Applicable
Electrical	________	________
Plant Steam	________	________
Gasses	________	________
Oil Free Compressed Air	________	________
Compressed Air	________	________
Utility Heated Water	________	________
Cooling Water	________	________
Potable Water	________	________
Purified Water	________	________
Clean Steam	________	________
House Vacuum	________	________

2.3 Operation and Maintenance Program

Complete data sheets in ATTACHMENT I including:
- Equipment Operation Documents
- Equipment Maintenance Documents
- Equipment Sanitation Documents

2.4 Instrument Calibration Program

Complete data sheets in ATTACHMENT I including:
- Critical Process Instruments
- Not for Calibration Instruments

2.5 Lubricants

Complete data sheets in ATTACHMENT I including:
- In Product/Critical Component Surface Contact
- Non In Product/Critical Component Surface Contact

2.6 Engineering Drawings

Complete data sheets in ATTACHMENT I including:
- Engineering Drawings

2.7 Controls, Alarms and Interlocks

Complete data sheets in ATTACHMENT I including:
- Controls, Alarms and Interlocks

2.8 Initial Cleaning

Complete data sheets in ATTACHMENT I including:
- Initial Cleaning

ATTACHMENT I
IQ Data Sheets

ATTACHMENT I
IQ Data Sheets – General
Equipment and System Description

Identify and document all equipment associated with the system undergoing IQ. Add additional pages if necessary. Make more tables like the following for all associated equipment.

Manufacturer	
I. D. Number	
Serial Number	
Model Number	
Location	

Work Orders
Identify all work orders associated with the system.

Completed By/Department Name	WO #	Description

Prepared By: ______________________ Date: ______________________

Reviewed By: ______________________ Date: ______________________

ATTACHMENT I
Manufacturers Specifications

Document where specifications associated with the system being validated documents are filed. Description	Location

Materials in Product Contact

List all materials in the equipment that come into product contact:

Part	Material

Prepared By: ______________________ Date: ______________________

Reviewed By: ______________________ Date: ______________________

IQ Data Sheets - Equipment and System Utilities
Electrical

Identify and document the installation's electrical requirements and the specific connections to the facility electrical supply.

Equipment Description

Service Requirements Actual Supply

Voltage		
Phase		
Amperage		
Frequency		

Safety Features

Circuit Breaker Installed	
Amperage Rating	
Circuit Breaker Location	
Circuit Breaker and Controls Properly Identified	
Other Electrical Safety Devices	

Supply Wiring Requirements Actual

Conductor Type		
Conductor Size		
Properly Grounded		

Prepared By: ______________________ Date: ______________________

Reviewed By: ______________________ Date: ______________________

ATTACHMENT I

IQ Data Sheets – Equipment and System Utilities
Gasses

Describe the source and connections of gasses used in manufacturing processes to the installation.

Source System Description

System Description	
Gas Type and Specification	
Validation Study ID (If Validated)	
Piping Material	
Supply Valve ID	

Required Supply Characteristics

Flow Rate or Pipe Size Specified	
Pressure Range Specified	

Actual Supply Characteristics

Flow Rate or Pipe Size	
Pressure Range	

Local Drop

Piping Material	
Dead Legs	

Filters

Filter Type	Manufacturer	Model	Porosity	Location Bldg./Room	Change Frequency

Make more tables on next pages with one each for the following system utilities since the description for these systems will remain system.

Plant Steam
Oil Free Compressed Air
Compressed Air
Potable or Utility Water
Chilled Water
Clean Steam
House Vacuum

Prepared By: ______________________ Date: ______________________

Reviewed By: ______________________ Date: ______________________

ATTACHMENT I

IQ Data Sheets – Equipment and System Utilities
Purified Water

Describe the source water system and its connections to the installation.
Source System

System Description	
Validation Study ID (If Applicable)	
PW Type (WFI, USP, DI, RO, etc.)	
Hot or Cold Supply/Supply Temperature	
Supply Piping Material	
Most Recent Date Tested and Results	
Supply Valve ID and Location	

Required Supply Characteristics

Flow Rate or Pipe Size Specified	
Pressure Specified	
Temperature Range Specified	

Actual Supply Characteristics

Flow Rate or Pipe Size	
Pressure	
Temperature Range	

Local Drop

Piping Material	
Dead Legs	

Prepared By: ______________________ Date: ______________________

Reviewed By: ______________________ Date: ______________________

ATTACHMENT I

IQ Data Sheets - Operation and Maintenance Program
Equipment Operation Documents

List SOPs, manuals and logs that are associated with operation of the system.

Document Number	Document Title	Revision

Make similar tables to list the SOPs, manuals and logs associated with Maintenance of the system and with cleaning, sanitation and sterilization of the system.

Prepared By: ______________________ Date: ______________________

Reviewed By: ______________________ Date: ______________________

ATTACHMENT I

IQ Data Sheets - Instrument Calibration Program
Critical Process Instruments

List all process instrumentation requiring calibration.

Instrument Type	
Manufacturer	
Serial Number	
Model Number	
Equipment Tag ID	
Calibration Tag ID	
Calibration Frequency	
Pre-OQ Calibration Date	
Post OQ Calibration Date	
Instrument Type	
Manufacturer	
Serial Number	
Model Number	
Equipment Tag ID	
Calibration Tag ID	
Calibration Frequency	
Pre-OQ Calibration Date	
Post OQ Calibration Date	

Prepared By: ______________________ Date: ______________________

Reviewed By: ______________________ Date: ______________________

ATTACHMENT I

IQ Data Sheets - Instrument Calibration Program
Not for Calibration Instrumentation

List all process instrumentation which does not require calibration.

Type	Manufacturer / Model

IQ Data Sheets - Lubricants
In Product/Critical Component Surface Contact

List all lubricants that come in product contact or in contact with critical components.

Name	Description	Used For

Prepared By: ____________________ Date: ____________________

Reviewed By: ____________________ Date: ____________________

ATTACHMENT I

IQ Data Sheets – Lubricants
Not In Product/Critical Component Surface Contact

List lubricants that do not come in product contact.

Name	Description	Used For

Prepared By: ______________________ Date: ______________________

Reviewed By: ______________________ Date: ______________________

ATTACHMENT I

IQ Data Sheets
Engineering Drawings

Identify engineering drawings that are associated with the equipment being validated

Drawing # and Storage Location	Title	Rev. #	Date

Controls, Alarms and Interlocks

Identify controls, alarms and interlocks that are associated with the item being validated.

Enter the name of the control, alarm or interlock in "Name", describe what it does in "Function Description" and specifically identify its physical location in "Location". Include asset numbers or instrument IDs in "Name".

Name	Function Description	Location

Prepared By: ____________________ Date: ____________________

Reviewed By: ____________________ Date: ____________________

ATTACHMENT I

IQ Data Sheets

Initial Cleaning

Describe and document all cleaning activities required to make this item ready for service.

Prepared By: ____________________ Date: ____________________

Reviewed By: ____________________ Date: ____________________

1. Initiate the OQ Protocol for a manufacturing equipment.

Answer:

Title	
Purpose	
Equipment	
Process/Product	
Component	
Location	
Department	
Documents Included	

Completion of the following signature block signifies that the appropriate parties have reviewed this validation protocol and agree with the methodology and acceptance criteria. A complete report with will be issued for separate approval.

Protocol Author: ____________________ Date: __________

Name/Title	Signature	Date
Manager Validation		
Manager Engineering		
Manager Quality Operations		
Director Quality Assurance		

TABLE OF CONTENTS

Title Page
Table of Contents

1 Introduction
1.1 Background
1.2 Purpose
1.3 Scope
1.4 References
1.5 Exceptional Conditions and Deviations
1.6 Validation Approach

2 Equipment and Operations
2.1 Process Equipment Studied
2.2 Critical Parameters
2.3 Critical Results
2.4 Apparatus, Equipment and Raw Materials Required
2.5 Calibrations

3 Responsibilities
3.1 Validation
3.2 Operations
3.3 Engineering
3.4 Quality Assurance
3.5 Regulatory Compliance
3.6 Technical Development

4 Sample Handling and Laboratory Results
4.1 Sample Identification and Labels
4.2 Laboratory Requirements
4.3 General

5 Evaluations
5.1 Evaluation 1
Attachments
Attachment 1- Sample Evaluation Summary
Attachment 2- Resolution of Problem Form

1 INTRODUCTION

1.1 Background

The reader needs to know what is being qualified, how it will be used, why it was selected for this service and where it will be used. One or two paragraphs should suffice.

This description may include (but is not limited to):

- *Equipment name and model*
- *Basic capacity and ratings*
- *Fixed or portable operation*
- *New or used equipment (with previous use)*
- *Planned use*
- *Why this item was selected for this service and why we feel that it will meet our needs*
- *Why the item requires validation*

This section should include a basic description of the equipment and its intended use. Like:

- *Electric pump to transfer water in manufacturing.*
- *Jacketed stainless steel pressure vessel for boiling water in manufacturing process.*
- *Clean steam generator to convert Purified Water, USP into clean steam using plant steam as the heat source. Attached to fluid bed dryers.*

1.2 Purpose

Describe the purpose of the validation program, why it is required and how it will affect the business.

1.3 Scope

Describe where this work applies (like a specific facility) or a set of specific conditions (like when a specific process condition occurs). Precisely define the equipment, system, process or product affected by the work

1.4 References

Validation SOP# ABC123 "Validation Program"

List appropriate local policies, SOPs, procedures or other documents that are affected by this study or the item being studied.

1.5 Exceptional Conditions and Deviations

Any minor changes or deviations that are necessary to comply with this protocol will be documented in the final report. Any significant changes or deviations altering the scope or intent of this protocol require prior approval by the original signatories and documentation in the final report.

This section insures that there is a mechanism for handling unforeseen events. It permits the study to continue if minor changes in the validation program or acceptance criteria are required.

1.6 Validation Approach

Describe the basic approach to validating this item. Mention general procedures, evaluations and methods that will be employed. Explain how this approach characterizes the operation of the item.

2 EQUIPMENT AND OPERATIONS

Specifically define all items affected by and included in the validation.

2.1 Process Equipment Studied

2.2 Critical Parameters

This is a subset of all the process control parameters. It includes only the parameters that are directly responsible for product quality. For example, it might include mixing time, air temperature or machine speed. It might not include line voltage or room positive pressure differential.

2.3 Apparatus, Equipment and Raw Materials Required

List all items required to perform the validation program other than the equipment being studied. This list would include: validation instruments, packaging components and process raw materials.

2.4 Calibrations

The following items require calibration to support this study:

Process Equipment
Process Instruments
Validation Instruments

Note any and all calibration requirements. Process equipment, process instrumentation, validation equipment, validation instruments, laboratory instruments, as required.

Calibration certificates or other evidence that the instruments were in calibration at beginning and end of this study will be provided in the validation report.

3 RESPONSIBILITIES

List all departments that will be involved in the validation program and specifically describe their responsibilities.

3.1 Validation

Prepare validation protocol

Perform the OQ study and prepare the final report

Review and approve validation protocol and report

3.2 Operations

Review and approve validation protocol and final report

Provide trained operators and other required resources to properly run necessary equipment

3.3 Engineering

Review and approve validation protocol and final report

Provide engineering and design specifications for the associated systems and equipment

3.4 Quality Assurance

Review and approve validation protocol and final report

3.5 Regulatory Compliance

Review and approve validation protocol and final report

3.6 Technical Development

Provide laboratory services for evaluation samples

Review and approve validation protocol and final report

4 SAMPLE HANDLING AND LABORATORY RESULTS

Describe how samples will be identified and labeled. If necessary describe specifically which laboratories will handle which samples. Describe any on routine sample handling requirements like "Hold under refrigeration or up to 24 hours".

4.1 Sample Identification and Labels

Each evaluation sample is assigned a unique identification number. The sample is described on the "Sample Evaluation Summary" in Attachment 1. The sample ID and evaluations required are transferred to the sample's label and recorded by the laboratory staff in their laboratory notebooks with the evaluation results.

The person taking the sample is responsible for; properly completing the "Sample Evaluation Summary"; preparation of a corresponding label; attaching the label to the sample and sample submission to the laboratory.

4.2 Laboratory Requirements

Perform all analysis per approved SOPs and validated analytical methods. Record results in laboratory notebooks referenced to Sample ID and Study Number. Provide copies of laboratory notebook pages or an official summary form to validation.

Describe how laboratory results will be recorded and reported. Specific forms may be referenced in the Evaluations section.

4.3 General

All data sheets and test results will be signed by appropriate validation personnel, analytical personnel and or process operators involved in this study, as required.

5 EVALUATIONS

Describe the planned evaluations and associated acceptance criteria.

5.1 Evaluation 1

Objective

Describe the reason this evaluation is being performed and what it reveals about the item being evaluated.

The way the evaluation will be performed is described in Methodology.

Methodology

Describe how samples will be taken including the sampling interval and where in the process samples will be taken from.

Identify specific evaluation methods either via an approved procedure or describe the procedure in full detail. If required describe how to dispose of samples afterwards.

Reference data sheets in Attachments if they are required.

Acceptance Criteria

List the specific acceptance criteria for this evaluation.

ATTACHMENT 1
SAMPLE EVALUATION SUMMARY

Sample Evaluation Summary

Sample Description	Sample ID	
	Batch ID	
	Drum ID	
Evaluation or Assay Required	Sample Date	
	Sample Time	
	Sampled By	
Sample Location	Acceptance Criteria	
Comments	Lab Results	
Sample Description	Sample ID	
	Batch ID	
	Drum ID	
Evaluation or Assay Required	Sample Date	
	Sample Time	
	Sampled By	
Sample Location	Acceptance Criteria	
Comments	Lab Results	

Data By: ______________________ Date: ______________________

Review By: ______________________ Date: ______________________

ATTACHMENT 2
RESOLUTION OF PROBLEM FORM

IN CASE OF A PROBLEM ARISES, THIS FORM WILL BE USED TO REPORT THE DETAILS OF THE PROBLEM, INVESTIGATIONS, ROOT CAUSE AND CORRECTIVE AND PREVENTATIVE ACTIONS

Report Number:
Problem Description:
Problem Reported by:
Problem investigations:
Resolution Plan:
Resolution Assigned to:
Resolution Approved for Implementation by:
Change Verified by:

Acceptance

Development Approval ______________________________ Date ______________________________

Validation Approval ______________________________ Date ______________________________

10.1 User Requirement Specification

Let us first discuss what a User Requirement Specification (URS) actually is? In Pharmaceutical Industry, before buying an equipment or a system, planning document is prepared that specifies what the equipment or system needs to do. This document is called URS. It is written from the point of view of the end user and does not need to be technical or complicated. The URS document contains requirements from multidisciplinary sources and supports design, commissioning and qualification activities, operations, and maintenance of the new system.

In the following pages, we present here a template or example of the URS using which as a guideline, the end user or an Engineering, Manufacturing, or Quality personnel can initiate a URS based on the system intended to be purchased. Where possible, the users should identify the source (e.g., studies, standards, etc.) for the acceptable ranges of variables or other critical requirements that have been derived.

USER REQUIREMENT SPECIFICATIONS FOR A SYSTEM

(INSERT YOUR SYSTEM DESCRIPTION HERE)

Table of Contents

1.0 Glossary 161
2.0 Introduction 162
3.0 Overview 163
4.0 Operational Requirements 163
  4.1 Capacity 163
  4.2 Process Requirements 163
  4.3 Process Control 164
  4.4 Functions 164
  4.5 Data and Security 166
  4.6 Environment 167
5.0 Constraint 168
  5.1 Equipment Constraints 168
  5.2 Compatibility and Support 169
  5.3 Availability 170
  5.4 Procedural Constraints 171
  5.5 Maintenance 171
6.0 Life Cycle 172
  6.1 Development 172
  6.2 Testing 172

DOI: 10.1201/9781003366393-14

6.3 Delivery 172
6.4 Support 173
7.0 References 174
8.0 Approval 174

1.0 GLOSSARY

If necessary, attach a glossary, which includes definitions of terms that may be unfamiliar to the supplier or terminology that may have meanings specific to entries on this User Requirement Check sheet.
For example (*example list is not intended to be complete*):

Acronym	Definition
°C	Degrees Celsius
B L	Biohazard Level
C F R	Code of Federal Regulations
d B	Decibels
E M I	Electro-Magnetic Interference
ft	Feet
G U I	Graphic User Interface
H M I	Human – Machine Interface
H V A C	Heating, Ventilation, and Air Conditioning
I S O	International Organization for Standardization

J E T T	Joint Equipment Transition Team
L E L	Lower Explosion Limit
O I P	Operator-Interface Panel
R F I	Radio Frequency Interference
U R S	User Requirement Specification

Project No.:
Insert the unique project number associated with this particular URS.
Document No.:
Insert the Document Identification Number and Revision.
Document Description:
Insert the description of document, e.g., PURE STEAM GENERATOR.

2.0 INTRODUCTION

Note: Insert your own description and introduction. You may wish to include your own requirements and specifics to allow the inexperienced reader, contractors, and suppliers adequate explanation to understand the scope of the equipment.

Describe who produced the document, under what authority, and for what purpose.

Describe the contractual status of the document. Include or reference as an attachment the "legal" paragraph that communicates purchasing terms and conditions in this section.

Describe the relationship of this document with other documents. This is important for equipment that is part of an integrated process or line and will help the Supplier to understand and ask questions that may otherwise be overlooked.

Refer to the Equipment Validation Plan that outlines the supplier's responsibilities attached.

Remove any options that are not applicable for this project, if other uses will apply list each separately.

This is to be utilized as a guide for the user to answer the majority of questions involved in specifying the "what" for the equipment. Addendums may be used to round out the requirements.

3.0 OVERVIEW

This document encompasses the normal range of equipment operations. This section should be expanded to provide a high level description of the system explaining why it is required and what is required of it. Include the background, key objectives, and the main functions and interfaces. Obviously, technology improvements and new applications may require deviation from this template.

The User should consider including information in the following:

- Use: Describe what applications the equipment/system will be used to support. Include a brief explanation of the general functions.

For example (for a wide range filler):

The wide range filler line shall process pharmaceutical vials under sanitary conditions with the following functions: vial filling, put stoppers to vials, oxygen reduction, sample check weighing, and vial eject.

For equipment dealing with solid dosage pharmaceutical products directly:

The products being developed on the specified equipment are of:

- Low active (vitamins and supplements)
- Medium active (e.g., Aspirin)
- Highly active (e.g., Morphine)
- Exclusive (e.g., Penicillin)
- Capacity: Provide a brief description of the equipment/system capacity requirements.

For example (for a wide range filler):

The equipment shall be capable of producing ABC product continuously, over an 8-hour shift.

4.0 OPERATIONAL REQUIREMENTS

4.1 Capacity

- Range of products
- Commodities to be used (i.e., vials, bottles, tablets, etc.)
- Processed materials to be used or involved
- Production rates

4.2 Process Requirements

- Quality of products and concentrations
- Product limitations (i.e., temperature, humidity, pH, etc.)

4.3 Process Control

- Measurement range
- Setpoint range
- Parameter control tolerance

Note: At a minimum, list the critical, process parameters. Consider documenting how each of the parameter requirements was determined.

4.4 Functions

Operation

Enter your specific operational requirements here.

- Cycles expected the system will operate at are…
- Recipes the system will contain are…
- Operator interaction with the system will include…

The following is an example:

The system shall operate with a minimum of operator involvement. Operation shall be safe from an operator and environmental standpoint.

The system shall be operated locally at the equipment itself.

Control system requirements:

- Manual only
- Manual/automatic

Power Failure/Recovery

Power Failure: Insert your specific power failure/recovery scheme here.

For example:

In the event of a power failure, the system shall protect in the following priority:

- Personnel
- Equipment
- Product

Recovery Methods: Insert your specific recovery scheme here.

For example:

- Manually restart based on operator inputs. (Operator initiates a power recovery sequence).
- Restart based on last state before loss of power. (System runs through a power recovery sequence when an operator start signal is given)
- If there is an automatic batch reporting system, the information shall be retained in the event of a power failure. (A UPS is required for automatic batch reporting systems for data retention).

"Emergency Stop"

Insert your specific "Emergency Stop" (E-Stop) strategy and communication scheme here.

For example:

The system shall have an E-Stop mechanism designed to stop all physical movement of the equipment immediately. The E-Stop mechanism(s) shall be located in easily accessible areas around the equipment as required by national and local safety standards.

Alarms and Warnings

"Critical alarms" – Insert your specific "Critical Alarm" action scheme here.

For example:

Critical Alarms shall take action via interlock(s) and/or operator procedural response to shut the equipment down and notify the operator of the condition(s). The operator shall be required to acknowledge the alarm before the alarm can be reset and the system restarted. Once the alarm is reset, the operator may restart the system.

"Informational Messages" shall notify the operator and take no further action.

Note: For a customized equipment application, include an alarm table, like the example that follows.

For example:

The equipment shall have the following critical alarms and warnings:

Alarm or Informational Message	Critical Alarm	Non-Critical Alarm	Informational Message	Response	
				Interlock(s)	Operator Procedural
Emergency Stop	X	X		X	
Control Platform Communication Watchdog	X			X	
Control Power Fault	X			X	
Main Air Fault	X			X	
XXXX		X			X
YYYY		X			X
ZZZZ		X		X	
Motor Faults	X			X	
CONTROL PLATFORM Battery Low Warning			X		

The list of critical alarms in the table is not intended to be a comprehensive list of all alarms for the system. A complete alarm list will be developed and included in the Functional Specification document.

For a standard equipment type application, request the supplier to submit the complete alarm list and differentiate between critical and non-critical alarms.

Note: Consider documenting how "critical and non-critical" criterions were determined.

4.5 Data and Security

Controls provided with data collection systems intended for use in the manufacture of pharmaceutical products shall be required to meet 21 CFR Part 11 compliance. The supplier shall stipulate the methods by which this criterion is met.

User Interfaces

This section should address the following issues:

- User interface access levels
- Frequency of data point collection
- Hardcopy/electronic data collection requirements
- Compliance with 21CFR Part 11 (for Europe – EREC Guidelines)
- Data retention time on the system
- Data storage media

Access to all Input/Output values and system status bits shall be provided through a data-communication link. Security for data and operator access is provided by (User ID/Password, Card Reader, Biometric (fingerprint, eye scan, voice recognition, etc.), etc.).

User Interface with Supervisors and Operators

Insert your interface scheme here. For example:

The CONTROL PLATFORM system shall include interfaces with the Operator and Supervisor that ensures easy, safe, and reliable operation.

An operator-interface panel shall be provided and mounted near the equipment or on the equipment. This panel shall provide the necessary switches, indicators, and devices to operate the equipment.

Interface with Other Equipment

The control system shall include the interfaces necessary to facilitate operation and configuration. For example:

- An RS-232 communications port shall be provided.
- A Modem communications port shall be provided.
- A high-speed configuration/monitoring connection shall be provided.

Security Levels

List the total number of different access levels that will be required. Provide a general description of the access rights for each level (screen navigation, operational control, control loop variable manipulation, alarm setpoint manipulation, etc.).

Data Collection

Consider any archiving and reporting requirements here.

- None
- Recorder
- Process printout
- Electronic process printout, historical trending
- Electronic process printout, historical trending and interface to company network

The following shall be recorded:

- XXXXX
- YYYYY
- ZZZZZ

4.6 Environment

Provide details of the physical environment in which the [equipment/system] will be operated.

Layout

For example:

- Allocated floor space for the equipment is _____ inches by ________ inches with at least a _______inch corridor around the periphery for the equipment.
- Vertical clearance is _________ inches.
- See attached drawing #_________________________ (if applicable).

Layout is specific to the size equipment being used and system requirements of the equipment.

Physical Conditions

Room Explosion Classification

List the critical explosion requirements for the room.

- Room will be non-hazardous
- Room will be Class I Div II (Zone 2) (explosive atmosphere present at known times)
- Room will be Class I Div I (Zone 1) (explosive atmosphere present at all times)
- Environmental Classification
- List the critical air quality requirements for the environment
- Class 100, 10,000, or 100,000 (*Federal Standard 209E*)

- ISO Class 5, 7, or 8 (ISO 14644-1 and ISO 14644-2)
- Class A, B, C, or D (The Rules Governing Medicinal Products in the European Union – Annex 1)

    - Directional airflow
    - Airflow velocity
    - Number of air changes
    - Particle ingress
        - Biohazard level
        - List the critical biohazard level requirements for the environment
        - Rating of enclosures
        - List the rating requirements for enclosures in the environment
            - NEMA 1 (Open)
            - NEMA 12 (Oil tight)
            - NEMA 4 (Wash down)
            - NEMA 4X (Wash down and non-Corrosive)

Cleaning Requirements

- Hand-Wash
- Auto-Wash
- Steam Clean

Sterilization Requirements

Describe the type of sterilization that will be performed on or within the [equipment/system] and the duration the [equipment/system] will be exposed.

Intended Operating Environment

Describe the type of operating environment that the [equipment/system] will be exposed to the following installation.

For example:

The [equipment/system] shall be mounted in a GMP environment with a temperature range of 15°C–25°C, non-condensing humidity.

Vibration levels in the operating environment are <Negligible, Slight, or Severe>

The Seismic Zone in the operating environment is <Zone 1, Zone 2, Zone 3, Zone 4, or Zone 4A>.

5.0 CONSTRAINT

5.1 Equipment Constraints

Describe the operating constraints that the [equipment/system] will be required to meet.

For example:

Environmental Conditions

The [equipment/system] is to be used under the following environmental conditions:

Altitude ________ ft above the sea level

HVAC Equipment Constraints

Process air enters in from...

Indoor Temperature: °C

Outdoor Temperature:

Minimum: °C

Maximum: °C

Climatic conditions:

Winter dry bulb °C

Winter wet bulb °C

Summer dry bulb °C

Summer wet bulb °C

Equipment Generated Vibration

The maximum allowable equipment vibration generated during operation is *<value>*.

5.2 Compatibility and Support

Include any requirements for compatibility with other user systems. In the examples provided below, fill in blanks with the appropriate information. Delete areas that are not applicable to this application.

Control Platform Controllers

The Supplier shall utilize Control Platform Controllers that shall include a communication port.

The Supplier shall provide documentation that the program (embedded software) was developed and coded utilizing program development and documentation software.

The supplier shall indicate the use of custom or "bespoke" code within the equipment, and shall supply licensing of the custom code for the life of the equipment.

An escrow agreement shall be required in the event the supplier cannot support the supplied custom software any longer. In such instances, the source code and all information regarding the source code shall be given to the user in a timely fashion to prevent the downtime of the equipment.

Preferred Vendor List

If necessary, attach any critical preferred vendors. Refrain from driving the supplier into customizing his equipment package.

If the vendor wishes to deviate from the Preferred Instrumentation List, a listing of the proposed components and instrumentation sub-vendors indicating type and model numbers shall be submitted with the quotation, along with an explanation of why the deviation is recommended.

Utilities

The utilities and space involved need to be discussed prior to purchasing the equipment. Obviously, the actual equipment size is but a small part of the entire scope of work necessary to implement a larger system. The unit and its support equipment must be able to be installed in current building facilities.

The User shall ensure that the following utilities are available and that the utility supply lines and piping are terminated with fittings or connections, which are compatible with those described on the Customer Connection Drawings.

The Supplier shall specify utility data, which are marked with a triangle. Utility data, which are not specified or marked with a triangle, shall be brought to the attention of the User. These data shall be specified (by the User or the Supplier) and shall be approved by both the User and the Supplier before system design begins.

Base Utilities Worksheet

- Related Discharges
- Volume
- pH
- Temperature
- Materials
- Electricity
- ______ VAC ____ Hertz
- Phases
- Amperage
- Uninterruptible Power Supply (Define Use)
- Control Air
- Pressure

For example:

_______ PSIG (must be clean, dry ISA instrument quality air, and capacity to be determined from the size of equipment). Recommend at least 90 PSIG constant air pressure for all equipment, unless specified.

Other Utility Requirements

Provide details regarding any other utility systems that are required to support operation of the equipment/system.

5.3 Availability

The [equipment/system] is intended to be operated [Continuously, Regularly, __ hours per day, __ hours per week].

Operation of the [equipment/system] shall be suspended, and the system shall be available for preventative maintenance or routine service [__ hours per week, __ continuous hours per week, __ days per month, __ weeks per year].

5.4 Procedural Constraints

Product Contact Materials

Describe any local procedural requirements such as local standards, which must be adhered to if applicable.

All piping welds and product contact welds shall meet ASME and 3A specification requirements. Product contact metal to be (316L Stainless Steel, 316 Stainless Steel, 304 Stainless Steel, etc.)

List critical requirements for surface finish, roughness, etc.

Other standards are attached. The Supplier shall adhere to the indicated sections only.

Product in Contact with Materials

All product contact parts should be considered. The [equipment/system] will be used for (aqueous solutions, low LEL (lower explosion limit) solvents, high LEL solvents, etc.).

Noise Level Constraints

Insert your own noise level constraint requirements here.

For example:

The noise level generated during operation of the [*equipment/system*] will not exceed 85 dB from 3 feet away.

EMI/RFI Levels Are

Insert your own electrical and radio noise immunity requirements here.

For example:

The [equipment/system] [will/will not] be used in an area where two-way radio communication devices are in operation. The [equipment/system] shall meet CE requirements for electrical noise immunity and emissions.

Containment

Explosion Protection

This section is only valid if the product being used is highly reactive. Insert containment guidelines here (i.e., Bar construction with isolation and suppression, 10 Bar construction with isolation only, Closed-Loop Nitrogen System, etc.).

Labeling

All equipment and control wiring shall be labeled and identified.

5.5 Maintenance

Describe the expected hardware and software maintenance support including: ease of maintenance, expansion capability (software only), likely enhancements (software only), expected lifetime and long-term support.

The system shall be maintained on a schedule as indicated by the supplier. The supplier is to provide (at minimum) the following maintenance instructions:

1. All sub-systems provided (Maintenance and operation manuals of vendor equipment)
2. A comprehensive lubrication list and recommended lubrication schedule
3. A comprehensive recommended maintenance (regular recommended inspect ion intervals, wear points, recommended spare parts list)
4. Supplier shall supply ______ Copies of Operation, Installation, Maintenance, and de-commissioning manuals
5. Accessibility for operation and maintenance personnel.

6.0 LIFE CYCLE

6.1 Development

If S88 is to be applied to the equipment being acquired, it should be referenced in this section of the document.

The Supplier shall provide a Quality and Project Plan as part of their proposal. The Supplier shall have a quality system in place. Internal quality procedures shall be available for the User's review.

The Supplier shall provide a Project Manager for the project to provide a single communication point with the User.

The project shall utilize the GAMP methodology when developing the system and documentation.

6.2 Testing

Describe the Supplier testing requirements. Reference the Validation Test Plan, Factory Acceptance Test, special tests, etc. This section should also include required amount of demonstrated run time, any special materials necessary to complete testing, integration testing, etc.

In order to verify system performance, the User shall witness the execution of the Factory Acceptance Test procedures. The Supplier shall notify the User _______ weeks in advance of the start of this test.

The Factory Acceptance Test Specification shall be submitted to the User for review and approval prior to execution. A minimum of _______ weeks shall be allowed for the User to review and to comment and/or approve the Factory Acceptance Test Specification.

Refer to the Equipment Validation Plan for applicable procedures.

6.3 Delivery

The [equipment/system], with all options, equipment, and the documentation listed below, shall be delivered to the User's receiving dock.

Documentation

Installation, operation, and maintenance instruction documentation for the system shall be developed to a level that is comprehensible to a high school graduate.

The Supplier shall provide the documentation for preliminary review. The Supplier shall provide documentation reflecting the "as-built" condition with final delivery.

All final documents shall be shipped with transmittals that identify them as contractually required documents. All final documents and drawings shall reflect the "as-built" condition.

All documents shall be in the language of the destination country and supplied with hard copies and electronic versions supplied in the format identified for each document:

• Project Plan	
• User Requirement Specification	
• Functional Specification/Requirement	
• Design Specifications	
• Controls Test	
• Hardware Installation Test	
• Operational Test	
• Factory Acceptance Test	
• Operator, Maintenance and Service Manuals	
• Process and Instrumentation Diagram (P&ID)	
• Instrument Listing	
• Control Schematics	
• Control Panel Assembly Drawings	
• Equipment Assembly Drawings	
• Bill of Materials	
• Spare Parts List	
• Component Cut Sheets	
• CONTROL PLATFORM Program Printout and Disk File	
• OIP Configuration Printout and Disk File	

6.4 Support

Describe what support activities are required after acceptance. The paragraphs outlined below provide some areas for consideration.

Startup Support (list available options)
Training (list training options available)

Post-Startup Support (list post-startup support available)
Technical Support
Telephone (Voice or Modem)
Replacement Parts Availability List (Normal lead times shall be listed)
User Site Support

Preventative Maintenance (list maintenance contracts available)
System Improvements (supplier shall notify user of any improvements available on a regular basis)

7.0 REFERENCES

List references that were used in preparing this document or that provide additional details, such as:

- Equipment Validation Plan
- Vendor Piping Certification (if applicable)
- Customer Connection Diagram
- Approved Instrumentation List
- Process Flow Diagram
- 21 CFR Part 11
- National Electric Code
- S88

8.0 APPROVAL

Insert your company's standard approval page at the appropriate position in the document or utilize the sample approvals listed below. The approval process/ requirements should be outlined in the Equipment Validation Plan.

This document has been reviewed by the User Project Manager and approved for use by the Supplier.				
Printed/Typed Name		**Signature**		**Date**
This document has been reviewed by the User Quality Assurance Group and approved for use by the Supplier.				
Printed/Typed Name		**Signature**		**Date**
This document has been reviewed by the Supplier Project Manager and approved for use by the Supplier Project Team.				
Printed/Typed Name		**Signature**		**Date**

Other approvals as necessary may be included, such as the Technical Services Groups, User Groups, Engineering, etc. as required.

In the following chapter, a URS of some particular equipment will be presented as an example for the readers.

10.2 Functional Specification

In the previous chapter, we have discussed about functional specification or FS. In response to a URS document, a separate document is prepared which is known as functional specifications. The functional specification is a kind of guideline and continuing reference point as the developers write the programming code. It is a formal document used to describe a product's intended capabilities, appearance, and interactions with users.

To understand the difference between a User Requirement Specification and the Functional Requirement Specification, consider this simple statement; User Requirements describe the end-user requirements for a system whereas the functional requirements describe what the system must do.

The reasons for writing functional specification documentation are:

1. The FS document serves as a roadmap for the developers of the system. Development is difficult enough with no structure or plan to guide them. Without having a functional specification document means spending lot more time on coding and producing code that's less efficient and of a lower quality and that will make it harder to fix bugs at a later time. One also risks ending up with an in-cohesive product that doesn't serve its purpose.
2. Another reason to use functional specification documents before designing product features is because functional specification documents are a way of getting everybody on board and dispelling personal assumptions about the product's features so that the features can be created that will solve real problems for the users. Once the process of requirement gathering and the functional specification has been established, you'll find that everything runs much more smoothly. Everyone will know what they're doing and will be operating from the same source of truth, with less back-and forth-between different departments.

Now the question is how to create a functional specification document?

There is no universally accepted functional requirements' document template, and it's up to the user which style and format to follow. However, there are several best practices that apply in most cases. There are a few steps that need to be taken:

1. Define the system's requirements.
2. Identify functionality required by the project.
3. Design how this functionality will work.
4. Describe the requirements and design in an understandable format.

DOI: 10.1201/9781003366393-15

Significant features of a FS document:

1. To make a good specification developer should ensure that all requirements are captured and all business rules are accurate.
2. Designers should define the user interface and interactions.
3. QA should be able to find enough information in the document to reflect the changes in tests.
4. Developers should find sufficient explanation to help them define technical specifications.
5. Technical writers should find answers to help them write meaningful user manuals and technical documentation.
6. Support teams should have a reference for qualifying customer reports as defects or normal behavior.

A functional specification is also a living document that may need to get updated over the entire lifecycle of the system. While it can and should usually be started ahead of development, leaving it along the way would be a serious mistake. The reason being that if the system is going to be kept for long and developers writing and maintaining code for thousands of customers over many years, a safety net is certainly needed when things go wrong. And if the specification is well written and up to date, having it will be of great significance that would save time.

In the following pages, an example is given as a template to write a functional specification. These are the information and details that a FS document should contain as a minimum requirement.

AUTHORS

Name	Role	Department

DOCUMENT HISTORY

Date	Version	Document Revision Description	Document Author

APPROVALS

Approval Date	Approved Version	Approver Role	Approver

Table of Contents

1 INTRODUCTION

Identify and describe the business need or problem this document will address. Include background information here.

1.1 Purpose of the Document

Describe what a functional specification document is and its intended purpose for the audience. Included below is a standard purpose of the FSD, but modify the description as needed.

The functional specification document is a document that provides detailed information on *how* the system solution will function and the requested behavior. This document is created based on the high-level requirements identified in the Business Requirements Document and provides traceability on the functional specifications back to the business requirements. Included in this document will be the detailed functional requirements including use cases, system inputs and outputs, process flows, diagrams, and mockups.

1.2 Scope

Describe the project scope that addresses the business need or problem. Include high-level information on the solution.

1.3 Related Documents

Add any related documentation that is relevant and related to the FSD. Some examples are the Project Charter, Business Requirements Document, etc.

Component	Name	Description

1.4 Terms/Definitions

State any terms and its definition that are described in the functional specifications. Include any acronyms that are mentioned in the document.

Term/Acronym	Definition	Description

1.5 Risks

List any assumed factors and identified risks that could affect the functional design of the system. Included can be third-party or commercial components that will be used, issues around the operating environment, or any constraints.

2 SYSTEM'S OVERVIEW

Provide a short description of the software and solution being specified and its purpose, including relevant benefits, objectives, and goals.

2.1 Context Diagram/Interface Diagram/Data Flow Diagram, Application Screen Flow, Sitemap, and Process Flow

Provide any appropriate graphical representations that are relevant to the system and project such as a context/interface/data flow diagram, application screen flow, site map, or process flow.

2.2 Roles and Responsibilities

2.2.1 User Roles and Responsibilities/Authority Requirements

User/Role	Frequency of Use	Security/Access, Features Used	Additional Notes
Include the specified user/role	Describes how often they use the system. State Frequent, Occasional or Rare	Describe the features of the system for the role and any security/ access permissions that should be stated	Add any additional notes or supporting documentation as necessary

2.3 Dependencies and Change Impacts

2.3.1 System Dependencies

List and identify any dependencies the proposed solution will have on other systems.

2.3.2 Change Impacts

List and identify existing systems that will be impacted by the implementation of the proposed solution.

3 FUNCTIONAL SPECIFICATIONS

In this section, describe the specifications related to the overall system here. You may want to create a table/ index of all functionalities explained in the sections below and link them to the items below

If no separate reference/ traceability document is created for the project, use this section to map the business requirements, use cases, functional requirements, and the test cases.

Group the functional specifications as appropriate for the project. May be divided by screens, functional areas, user role, high-level functions vs detailed functions, or any other way that works for your project.

3.1 <TITLE>

3.1.1 Purpose/Description

Include a high-level description and purpose of the specifications covered in the section.

3.1.2 Use Case

Map the functional requirement to one or more use cases mentioned in the Business Requirements Document. If the use case is not described in detail in the Business Requirements Document, describe the use case here. This typically includes the element s in the following table.

	<Use case name>
Primary Personnel(s)	*< that participate in this use case>*
Stakeholders and Interest	*< describing other stakeholders>*
Trigger	*<action that initiates/starts the use case>*
Pre-conditions	*<assumed to be true before the first step>*
Post-conditions	*<after the use case is successfully executed>*
Main Success Scenario	*1. STARTING-POINT* *2. Step* *3. Step* *4. GOAL-ACHIEVED>*
Extensions	*If Condition, then Alternative Steps* *List any extended steps/ scenarios that occur, other than the main success scenario.*
Priority	*high, medium or low*
Special Requirements	*<system related special requirements needed to fulfill the use case>*

3.1.3 Mock-Up

Provide the mock-up of the functionality or a mock-up of the entire page.

3.1.4 Functional Requirements

Describe the page level details. These may include any requirements related to Navigation Menu, Actions, transaction status, verification and validation requirements, etc.

Make sure each specification has a reference number and is explained in the following format.

Spec ID	Specification Description	Business Rules/Data Dependency
<Specification Identifier>	<Short explanation of the specification>	<Any validation rules or business rules>

4 SYSTEM CONFIGURATIONS

<Provide an overview of all the steps or the set ups required to configure an application/program. Also state the intent or purpose behind each set up or configuration. Discuss the possible alternatives, customizations, workaround's, conditions, and dependencies in a particular configuration. In case of Oracle applications, please list all applicable BR100's or Application set up documents>

5 OTHER SYSTEM REQUIREMENTS

<This section is used in contrast with stated functional requirements to highlight the additional details on the quality related aspects as well as other behavioral aspects of a system. This section is used to capture the stakeholders' implicit expectations about how well the system will work under a given circumstance. Here you can state the specific SLAs related to system response times (data search and retrieval), performance needs and metrics, latencies in a particular timeframe or during high-volume transactions, system failures and recovery management, security levels and accessibility constraints, data backup and archiving capabilities, legal compliance needs, etc. The broader definition of the term "system" also includes integrations with all types of mobile platforms, mobile devices, tablets, and smart phones.>

6 REPORTING REQUIREMENTS

<This section is used to capture the reporting needs, including but not limited to the scope and format of the report, data elements and contents required on the report, file types and extraction mechanisms, user base and accessibility levels, frequency of report extractions, etc. Also provide the mock-up of the report if needed. If necessary, create a separate document for reporting requirements.>

7 INTEGRATION REQUIREMENTS

<Identify the integration needs and state all required interfaces with anything external to this solution including hardware, software, and users. Include Architectural overview diagrams, high-level data flow diagrams, table structures and schema, interface protocols, API's, Error conditions, Error validations and messaging needs, Auto processing requirements, etc. You can optionally state hardware and software dependencies, Upgrade requirements, compatibility issues with existing frameworks and solutions, etc.>

(Data Flow Diagrams, Interface Diagrams – if necessary)

7.1 Exception Handling/Error Reporting

<This is where you can explain the error conditions/exceptions that normally happen in interfaces or cross flow system integrations. Explain the nature of exception, Error Id, Root cause of the error and also the strategy to handle the scenario. You can also indicate if there are any concurrent programs designed to automatically handle the error records or error conditions. State if there are any error reports generated or notifications utilized to alarm the support teams and system Administrators during the interface failures or outages.>

Exception/Error ID	Error	Cause	Solution Strategy

8 DATA MIGRATION/CONVERSION REQUIREMENTS

<Explain in brief the data conversion plan. Provide full identifying information for the automated system, application, or situation for which the data conversion plan applies. Describe briefly any assumptions, constraints or risks regarding the data conversion effort. (Provide details in Section 1.6.)

8.1 Data Conversion Strategy

<Include the overall strategy for the Data Conversion. This includes how and when you will perform the conversion – the approach used to extract, transform and load data during the conversion process, the conversion schedule, and test plan for testing the converted data.>

8.2 Data Conversion Preparation

<Provide details on any prerequisites necessary for the conversion. Discuss the backup strategy, restoration process in case the conversion fails.>

8.3 Data Conversion Specifications

Source	Source Data Element	Target	Target Data Element	Conversion Rules	Notes
<Source Location>	<Source Data Element Identifier >	<Target Location>	<Target Data Element Identifier>	<Describe Rules for Data Conversion>	<Additional Notes>

9 REFERENCES

<List all references to external material used as background information or knowledge for the FSD. Examples may include a compliancy website, Stanford website, etc.>

11 Stability Studies, Stability Indicating, and Methods Degradates

The stability program ensures product quality, safety, and efficacy meet throughout the shelf life. This is considered a prerequisite for the acceptance and approval of any pharmaceutical product. These studies must be conducted and planned following the guidelines issued by regulated authorities.

Stability testing provides evidence of how a drug substance or product's quality varies with time under the influence of various environmental factors such as temperature, humidity, and light. It also establishes a re-test period for the drug substance or shelf life for the drug product and recommended storage conditions.

In a stability program, products are kept under certain conditions in a chamber. This chamber is validated for temperature and humidity.

The stability studies are pulled on certain time points and tested. The documented stability proves that products are stable up to a certain time under a certain condition. Stability also provides the expiry of products.

STABILITY

Stability is defined as "The capacity of a drug product to remain within specifications established to ensure its identity, strength, quality, and purity".

EXPIRY DATE

The date placed on the container/labels of a drug product designates the time during which a batch of the product is expected to remain within approved stability specifications if stored under defined conditions.

SHELF LIFE

The time interval that a drug product is expected to remain within approved stability specification if it is stored under the conditions defined on the label in the proposed container/closure system.

ACCELERATED TESTING

Studies conducted under defined conditions for temperature and humidity are designed to increase the rate of chemical degradation or physical change of a drug

DOI: 10.1201/9781003366393-16

substance or drug product by using exaggerated storage conditions as part of the formal stability studies.

LONG TERM TESTING

Stability studies under the recommended storage conditions for the retest or shelf life proposed (or approved) for labelling.

BRACKETING

The design of the stability schedule so that at any time point, only the samples on the extremes (e.g., container size and/or dosage strengths) are tested. The design assumes that those at the extremes represent the stability of the intermediate condition samples. Where a range of dosage strengths is to be tested, bracketing designs may be particularly applicable if the strengths are very closely related in composition (e.g., tablet range made with different compression weights of a similar basic granulation, or a capsule range made by filling different plug fill weights of the same basic composition into different size capsule shells). Where a range of sizes of immediate containers is to be evaluated, bracketing designs may be applicable if the material composition of the container and type of closure are the same throughout the range.

MATRIXING

The statistical design of a stability schedule so that only a fraction of the total number of samples are tested at any specified test point. At a subsequent test point, different sets of samples of the total number would be tested. This design assumes that the stability of the samples tested represents the stability of all samples. In every case, all batches must be tested initially and at the end of the long-term testing (at expiry). The differences in the samples for the same drug product should be identified as, for example, covering different batches, different strengths, and different sizes of the same container/closure system and possibly, in some cases, different container/closure systems. Matrixing can cover reduced testing when more than one variable is being evaluated, and, hence, the design of the matrix will be dictated by the factors to be covered and evaluated. This plan works well for those products with moderate stability and small variability in the stability data. It is desirable to discuss the design with regulatory authorities in advance and the Matrixing proposal should be submitted in a pre-approval supplement.

ACCELERATED STUDIES

Accelerated studies are tested for up to six months.

Accelerated testing of the drug substance helps identify degradation products, which can help to establish the degradation pathways and validate the stability using the analytical procedures used.

Stress testing is likely to be carried out on a single batch of the drug substance. It should include the effect of temperatures (in 10°C increments (e.g., 50°C, 60°C, etc.)

above that for accelerated testing), humidity (e.g., 75% RH or greater) where appropriate, oxidation, and photolysis on the drug substance.

Examining degradation products under stress conditions is helpful in establishing degradation pathways and developing and validating suitable analytical procedures. However, it may not be necessary to examine specifically for certain degradation products if it has been demonstrated that they are not formed under accelerated or long-term storage conditions.

Results from these studies can be provided to regulatory authorities to register the product.

SELECTION OF BATCHES

Data from accelerated stability studies can be provided on at least three primary batches of the drug substance. The batches should be manufactured to a minimum pilot scale by the same procedure and using a method of manufacture and procedure that simulates the final process to be used for production batches. The quality of the batches of the drug substance placed on formal stability studies should represent the quality of the material to be made on a production scale.

CONTAINER CLOSURE SYSTEM

The stability studies should be conducted on the drug substance packaged in a container closure system that is the same as the packaging proposed for storage and sale.

TESTING FREQUENCY

For long-term studies, the frequency of testing should be sufficient to establish the stability profile of the drug substance. For drug substances with a proposed testing period of at least 12 months, the frequency of testing under the long-term storage condition should usually be every three months over the first year, every six months over the second year, and annually thereafter through the proposed re-test period.

Under the accelerated storage condition, a minimum of three-time points, including the initial and final time points (e.g., zero, three, and six months), from a six-month study is recommended.

If accelerated stability fails, intermediate storage studies are activated.

When testing under the intermediate storage condition is called for as a result of significant change under the accelerated storage condition, a minimum of four time points, including the initial and final time points (e.g., 0, 6, 9, 12 months), from a 12-month study is recommended.

STABILITY STORAGE CONDITIONS

The environment into which stability batches are placed for storage is known as Stability Storage Conditions. The conditions are controlled and monitored to maintain predetermined temperature, relative humidity, and/or light settings within allowable tolerances.

A documented plan to monitor and establish how the quality of a finished drug product may vary over time under the influence of environmental factors, including temperature, humidity, and light, is known as a stability program.

Generally, a drug substance should be evaluated under storage conditions (with appropriate tolerances) that test its thermal stability and, if applicable, its sensitivity to moisture. The storage conditions and the lengths of studies chosen should be sufficient to cover the storage, shipment, and subsequent use.

The long-term testing should cover a minimum of 12 months on at least three primary batches at the time of submission and should be continued for some time sufficient to cover the proposed re-test period. Data from the accelerated storage condition and, if appropriate, from the intermediate storage condition can be used to evaluate the effect of short-term excursions outside the label storage conditions (such as might occur during shipping).

STABILITY CHAMBERS

Stability samples are placed in stability chambers. Stability chambers are designed to maintain certain humidity and temperature.

The typical temperatures and humidity are as follows:

LONG-TERM STUDIES

25°C ± 2°C/60% RH ± 5% RH or 30°C ± 2°C/65% RH ± 5% RH, for 12 months.

INTERMEDIATE STUDIES

30°C ± 2°C/65% RH ± 5% RH, for six months.

Accelerated studies: 40°C ± 2°C/75% RH ± 5% RH, for six months

STABILITY COMMITMENT

When available long-term stability data on primary batches do not cover the proposed re-test period granted at the time of approval, a commitment should be made to continue the stability studies post-approval in order to firmly establish the re-test period.

Where the submission includes long-term stability data on three production batches covering the proposed re-test period, a post-approval commitment is considered unnecessary. Otherwise, one of the following commitments should be made:

If the submission includes data from stability studies on at least three production batches, a commitment should be made to continue these studies through the proposed re-test period.

If the submission includes data from stability studies on fewer than three production batches, a commitment should be made to continue these studies through the proposed re-test period and to place additional production batches, to a total of at least three, on long term stability studies through the proposed retest period.

If the submission does not include stability data on production batches, a commitment should be made to place the first three production batches on long-term stability studies through the proposed re-test period.

EVALUATION

The purpose of the stability study is to establish, based on testing a minimum of three batches of the drug substance and evaluating the stability information (including, as appropriate, results of the physical, chemical, biological, and microbiological tests), a re-test period applicable to all future batches of the drug substance manufactured under similar circumstances. The degree of variability of individual batches affects the confidence that a future production batch will remain within specification throughout the assigned re-test period.

The data may show so little degradation and so little variability that it is apparent from looking at the data that the requested re-test period will be granted. Under these circumstances, it is normally unnecessary to go through the formal statistical analysis; providing a justification for the omission should be sufficient.

An approach for analyzing the data on a quantitative attribute expected to change with time is to determine the time at which the 95% one-sided confidence limit for the mean curve intersects the acceptance criterion. If the analysis shows that the batch-to-batch variability is small, it is advantageous to combine the data into one overall estimate. This can be done by first applying appropriate statistical tests (e.g., p values for the level of significance of rejection of more than 0.25) to the slopes of the regression lines and zero-time intercepts for the individual batches. If it is inappropriate to combine data from several batches, the overall re-test period should be based on the minimum time a batch can be expected to remain within acceptance criteria.

The nature of any degradation relationship will determine whether the data should be transformed for linear regression analysis. Usually, the relationship can be represented by a linear, quadratic, or cubic function on an arithmetic or logarithmic scale. Statistical methods should be employed to test the goodness of fit of the data on all batches and combined batches (where appropriate) to the assumed degradation line or curve.

Limited extrapolating of the real-time data from the long-term storage condition beyond the observed range to extend the re-test period can be undertaken at approval time if justified. This justification should be based on what is known about the mechanism of degradation, the results of testing under accelerated conditions, the goodness of fit of any mathematical model, batch size, the existence of supporting stability data, etc. However, this extrapolation assumes that the same degradation relationship will continue to apply beyond the observed data.

Any evaluation should cover not only the assay but also the levels of degradation products and other appropriate attributes.

STATEMENTS/LABELING

A storage statement should be established for the labelling in accordance with relevant national/regional requirements. The statement should be based on the stability evaluation of the drug substance. Where applicable, specific instructions should be provided, particularly for drug substances that cannot tolerate freezing. The terms "ambient conditions" or "room temperature" should be avoided.

A re-test period should be derived from the stability information, and a retest date should be displayed on the container label if appropriate.

STABILITY PROTOCOL

A stability protocol is written and approved by management.

A typical stability protocol consists of the following sections.

- Product and strength.
- Packaging size
- Product manufacturer.
- Number of batches to be placed on study.
- Tests to be performed.
- Testing frequency.
- Storage conditions.
- Expiration Date.
- Details of the validated method.
- Acceptable/approved limits for each test.

STABILITY-INDICATING METHOD

A stability-indicating method (SIM method) can be defined as a "Validated quantitative analytical method that can detect the change with time in the chemical, physical or microbiological properties of the drug substance and drug products are specific so that the content of active ingredients and degradation products can be accurately measured without interference".

When analyzing a fresh product for release the product is fresh. With a period, the product starts to degrade. The active drug starts to convert into related compounds. So the SIM method must be capable of analyzing those degrades. Those compounds must be identified, their toxicity must be evaluated, and the limit of those related compounds must be specified.

The product is treated to some drastic conditions to break the active in related compounds.

This process is called force degradation or stress studies.

FORCE DEGRADATION

Forced degradation/stress testing is defined as "the stability testing of the drug substance and drug product under conditions exceeding those used for accelerated stability testing". Degradation can be achieved by exposing the drug, for an extended period, to extremes of pH (HCl or NaOH solutions of different strengths), at elevated temperature, to hydrogen peroxide at room temperature, to UV light, and to dry heat (in an oven) to achieve degradation to the extent of 5%–20%. Generally, trial and error experimentation are used during these experiments.

QUESTIONS AND ANSWERS

1. **Define stability?**
   Answer: Stability is defined as "The capacity of a drug product to remain within specifications established to ensure its identity, strength, quality, and purity".

2. **What is the expiry date?**
   Answer: The date placed on the container/labels of a drug product designates the time during which a batch of the product is expected to remain within approved stability specifications if stored under defined conditions.

3. **What are accelerated studies?**
   Answer: Studies conducted under defined conditions for temperature and humidity are designed to increase the rate of chemical degradation or physical change of a drug substance or drug product by using exaggerated storage conditions as part of the formal stability studies.

4. **What are the storage conditions for long term, intermediate and accelerated studies?**
   Answer: Long-term studies: 25°C ± 2°C/60% RH ± 5% RH or 30°C ± 2°C/65% RH ± 5% RH, for 12 months.
   Intermediate studies: 30°C ± 2°C/65% RH ± 5% RH, for six months.
   Accelerated studies: 40°C ± 2°C/75% RH ± 5% RH, for six months

12 Technology Transfer

The term "Technology Transfer" refers to the chain of processes that are needed for successful development of a drug product from discovery stage to full-scale commercial batch manufacturing.

This could further be explained as the transfer of the product from one technology company or site to another, to help make the product accessible to those area people where it is not available. The transfer of a drug product from Research and Development pilot batches to commercial production floor for a scale up batch is also called as "Technology Transfer".

A successful tech transfer not only requires knowledge about products and processes but also knowledge about equipment, procedures, competencies, roles and responsibilities, cultures, regulatory, and training requirements.

From a pharmaceutical manufacturer's perspective, the focus is on how to deliver better results to clients from a project management point of view.

There are key considerations to lead to a successful pharmaceutical tech transfer and ultimately a product launch to market including a dedicated project management program, standardized tech transfer procedures and robust quality systems.

HOW TO MANAGE TECHNOLOGY TRANSFER PROJECT

In order to carry out technology transfer effectively, there must be dedicated group who could be assigned the responsibilities to identify and fulfill the requirements prior to execution.

- The project management needs to work out with the business development team (BD) to understand the products/services or client needs.
- The team then assesses the technology package in collaboration with other internal departments.
- To understand the client's product manufacturing requirements, the project management team has to first understand the product and process requirements in detail and then evaluate its own capabilities to be able to successfully meet the needs.
- The next step is to generate a proposal and preliminary plan to successfully complete a technology transfer for the client's product and perform routine commercial manufacturing at their site. In case the technology transfer is being carried out in the same company from R&D to the production department, then the client would be the company's production department.
- After the proposal is approved by the client (or the production department) and the project is initiated, the project team will form a project management program to start the overall and thorough project planning before execution.

DOI: 10.1201/9781003366393-17

- Besides the project management team, a project manager could be the key personnel to contact and coordinate with for a successful and smooth technology transfer. The project manager is required to monitor and look after the transfer activities, track development, resolve issues, report milestones, and communicate effectively with the top management and other stake holders about the project.
- The Project Team should schedule periodic meetings to ensure proper communications on project progress, key decision makings, issue discussion, and risk management.

NECESSARY ACTIONS OF PROJECT TEAM FOR TECHNOLOGY TRANSFER

Before going into the technology transfer, a responsible project manager has to look into the following important points for completion:

1. The regulatory affairs department should first submit a detailed assessment package and got the submission file approved by the corresponding Health regulatory body.
2. An approved change control discussing and outlining all necessary steps and requirements is in place. The project manager must ensure that the documentation of changes was approved by the client so that the interpretation of discrepancies between expected and actual values could be possible. It is also advisable to form a system to track these changes during the transfer process.
3. Availability of the Master manufacturing formula for the scaled up batch.
4. Availability of approved raw materials and products test methods and specifications.
5. Sufficient quantity of raw materials and APIs for three or more validation batches. Sometimes unforeseen conditions or anomalies may lead a validation batch to failure and as a result the more quantity could be required to complete the three consecutive successful validation batches.
6. Availability of successfully qualified and like-to-like equipment to accommodate the larger size batches.
7. Analytical methods are transferred to the site where the scale up batch manufacturing will be carried out and are successfully validated. Analytical methods are needed in the very initial stages of technology transfer as all the studies require certain testing to provide data and demonstrate the study results. Raw materials that meet the specifications need to be sourced, purchased, tested, and released to support study runs and process qualifications.
8. Analytical test equipment are successfully qualified and available for use.
9. All necessary reference standards and reagents are available.
10. The process validation protocol is approved with all sampling and testing plans incorporated.

Besides all those above-mentioned needs, the presence of experienced and technically competent personnel is most important to accomplish the technology transfer batches successfully.

Once the above-mentioned requirements have been met, the site is considered fully ready to carry out the technology transfer process.

CHALLENGES IN TECHNOLOGY TRANSFER

Most of the pharmaceutical industries have a very robust technology that works smoothly to transfer the development batches to routine commercial scaled up batches successfully. However, there are always new challenges that may come across to surprise the personnel during the process. These challenges may be associated with the different product processes, equipment design and scale, regulatory needs, communication skills, and some other factors like geographic locations. Each company should set up its standardized procedure to manage technology transfers in a very structured and consistent way. A dedicated project management team is a plus to work closely with the project team to communicate and to track progress along the way to ensure timely and successful transfers.

QUESTIONS AND ANSWERS

1. **What is technology transfer?**
   Answer: The term "Technology Transfer" refers to the chain of processes that are needed for the transfer of the product from one technology company or site to another to help make the product accessible to those areas people where it is not available. The transfer of drug product from Research and Development pilot batches to commercial production floor for a scale up batch is also called as "Technology Transfer".

2. **What is the first step in technology transfer?**
   Answer: The first step in technology transfer is the submission of a detailed assessment package to get the file approved by the corresponding Health regulatory body.

3. **What is the rationale behind having an approved change control prior to commencement of technology transfer project?**
   Answer: An approved change control discussing and outlining all necessary steps and requirements is necessary prior to the start of the project so that the documentation of changes could be already approved and understood by the client and the interpretation of discrepancies between expected and actual values could be possible. The approved change control also helps to identify all the needs in time before the project commences.

4. **What is the strategy to be followed for the initial tech transfer batches manufacturing?**

   Answer: The first three batches in technology transfer are always made as validation batches. These batches are manufactured under an approved and agreed upon validation protocol with excessive sampling and testing plans that could challenge the robustness of the processes and various other factors involved in the tech transfer, for example, size of the batches, difference in equipment type, and or change in location including environmental conditions.

13 Dispensing

The Good Manufacturing Practices are to be practiced everywhere in a pharmaceutical industry. There is no area or process which is free from GMP and this includes even the dispensing process. The process of dispensing comprises of weighing and dispensing of specific amounts of raw material in powder or liquid form from the bulk container into smaller units that will be released and added in specific quantities for batch manufacture.

STEPS INVOLVED IN RAW/PACKAGING MATERIAL DISPENSING

Dispensing is where the manufacturing process of a particular product starts. It is one of the most critical steps as weighing and dispensing the correct amount of materials are very important in drug manufacturing. The weighing and dispensing of active pharmaceutical ingredient (API) is especially critical and any error in the weighing machine or a human error could lead to a serious and major non-compliance issue which could jeopardize the entire lot or could endanger patient's life. The reason being that the incorrect material by weight or selection of wrong material may not be detected by quality control testing. Therefore, quality assurance has to establish a system that could protect such mistakes and human errors right from the stage of dispensing.

The dispensing process comprises the following steps:

1. According to the Batch manufacturing, a bill of material (BOM) is raised by the production department and after necessary reviews and approval by Quality Assurance department, the BOM is sent to the Material management for dispensing the required amounts of material.
2. The responsible person in the store will review the material requisition sheet or the BOM and ensure the availability of all of the materials in the list.
3. In case a material is not released yet or still under QC test, the QC manager is communicated to expedite the release of the batch.
4. The Store person will check and release the containers of the required material on the FIFO basis.
5. All the released material shall have a RELEASED tag on the containers with all necessary information about the material name, code, batch number, release date, manufacturing and expiry dates, re-test date, etc. The label must have an approval by QA, signed off by the responsible QA personnel.
6. The containers will be taken to the dispensing booth where a store person will weigh and dispense the required quantity of the material in separate bags.
7. Each person entering the dispensing booth will follow the secondary gowning procedure.

DOI: 10.1201/9781003366393-18

8. Use required personal protective equipment (PPE) such as hand gloves, nose mask, Earmuff, Earplug, and safety goggle while working in dispensing area.
9. The cleanliness of the dispensing booth will be checked by a store person to ensure there is no container or material bags left from a previous dispensing.
10. Each and every weighing machine used in the dispensing booth shall have a calibration sticker on them clearly showing the calibration status of the balance, signed off by maintenance and checked and approved by QA personnel.
11. Only one material will be taken to the dispensing booth at a time to avoid a mixup.
12. Environmental conditions of the dispensing area will also be checked by the store person so that any temperature, relative humidity, or light sensitive material could be taken care off. To ensure the recommended environmental conditions, all the magnehelic gauges and instruments will be examined for their current readings as well as their calibration status.
13. The store person shall also check and verify the cleanliness and dryness of the scoops and other tools to avoid any cross contamination.
14. Once all these steps have been taken, the QA personnel will give clearance to the store to start the dispensing process for the specific batch.
15. All the materials will be dispensed in clean and dry polybags, immediately closed and tightened with a weighing slip.
16. The dispensing personnel will clearly indicate the name of the material, tare weight, gross weight, and net weight on the weighing slip, sign, and tie it up with the bag.
17. The QA personnel will check each and every material label for correctness and verify the weight written on it with the actual weight on the BOM.
18. In case of a discrepancy found, it will be escalated and brought into the notice of quality control, quality assurance, and store managers.
19. After the completion of dispensing process, the dispensed materials enclosed in bags will be locked in a secured cage to be transferred to the manufacturing floor.
20. The corresponding production supervisor or manager will receive the materials cage and will verify the materials weight and other details as peer the BOM and will give his approval for receiving the materials.

This will complete the dispensing process and the responsibilities of the stores personnel after QA verification and approval with each step.

IMPORTANT POINTS TO CONSIDER

The stores personnel will ensure:

- A separate scoop is used to dispense each material
- All scoops are cleaned and dried
- Material will be dispensed in double polybags

- Materials will be dispensed under a Laminar Flow hood
- Before dispensing active raw material, the potency will be re-checked and verified
- The dispensed materials shall be sent to the production floor through the passbox with the original and duplicate Bill of Materials
- The duplicate of BOM shall be used as receipt and will be signed off by the production personnel to be sent back to the store manager
- The last container from where the material was used will be marked with a label "In use" and the remaining quantity will also be put on the label
- After the dispensing is finished, stores person will clean the area to ensure no remaining traces of the materials weighed remain there and the Laminar Flow Hood will be switched off

DISPENSING OF PACKAGING MATERIALS

The primary and secondary packaging materials are also issued under the similar kind of checks and verification by stores and QA persons except for weighing and some other steps which may be required only for raw materials like usage of Laminar Flow or scoops which are obviously not required for packaging materials. The primary packaging materials (PPMs) are dispensed under a reverse Laminar Flow to clean any foreign particle lying on the body of the PPM.

The most important PPMs include:

- Vials.
- Bottles.
- Blister packs.
- Sachets.
- Syringes.
- Ampoules.

The PPMs are more critical as they have a direct contact with the products. That is why to check and verify the environmental conditions of dispensing booth is equally critical for issuance of PPMs.

Since the releasing of PPMs follows a similar procedure like raw materials, the necessary checks of the released lots and issuance of BOM for packaging materials, dispensing, and verification by QA personnel are also very critical and important to maintain the quality, efficacy, and safety of the product, the materials are being used for.

The area clearance of dispense booth from any previous PPM is also important here to ensure right materials with correct product's name and strength is being released for use and that no previous product documents or material is available in the area.

Store person must also ensure that differential pressure reading on the magnehelic gauge in the area and the HEPA filters is within the allowable limit for primary and secondary packaging materials.

Just like the raw materials, the PPMs shall also be dispensed in the cage/container having lock and key arrangement.

Labels shall be issued by counting them in numbers. If the labels are available as Roll labels, the entire Roll shall be given (if the Roll has a counter on the backside of the label).

DISPENSING OF PACKAGING MATERIALS FOR STERILE PRODUCTS

There are some rules to dispense packaging materials for sterile products. However these rules are just additional to the normal practice that we have seen earlier in the chapter for dispensing of raw and other packaging materials.

These rules are as follows:

1. Vial and rubber stopper can be issued in full. If the full quantity differs from the required quantity, a loose pack can be issued only if the material is not pre-sterilized.
2. Pre-sterilized material shall be issued in the manufacturer's original pack; it cannot be issued in loose quantity. While dispensing such material, the integrity of the pack needs to be ensured.
3. Flip of seals, needle guard, and plunger rod to be issued in full pack. If pack quantity is differing from the required quantity, it can be issued by weight. Standard average weight can be used to calculate the quantity.
4. After completing dispensing, the remaining material container shall be transferred to the warehouse at a designated place.

After completion of dispensing activity, personnel working in the area shall remove the hand gloves and cut them to prevent re-use of it. Discard the gloves in the waste bin.

QUESTIONS AND ANSWERS

1. **What is dispensing process in the pharmaceutical industry?**
   Answer: The process of dispensing comprises weighing and dispensing of specific amounts of raw material in powder or liquid form from the bulk container into smaller units that will be released and added in specific quantities for batch manufacture.

   The same approach is applied on dispensing of packaging materials where the materials are dispensed in numbers instead of weighed amount.

2. **Why is dispensing process considered critical?**
   Answer: Dispensing is one of the most critical steps as weighing and dispensing the correct amount of materials is very important in drug manufacturing. The weighing and dispensing of active pharmaceutical ingredient (API) is especially critical and any error in the weighing machine or a human error could lead to a serious and major non-compliance issue which could jeopardize the entire lot or could endanger patient's life.

Similarly, in case of primary packaging material, packs with correct information printed are critical while dispending as any incorrect information printed on the PPM or a mixup in the packaging material with contradictory information could be hazardous to the patient's health.

3. **Why the calibration status of the weighing machine in dispensing booth must be cross-checked by QA?**
   Answer: In case of a non-calibrated weighing machine, the amount of material weighed could be incorrect which may lead to non-conformance of the entire batch if noted during the QC tests. If not, then the efficacy, quality and safety of the product will be jeopardized and a serious hazard to patient's health could happen. Therefore it is important that the weighing machines should always stay in calibrated state.

4. **Why primary packaging materials are considered more critical then secondary?**
   Answer: The primary packaging materials (PPMs) are more critical as they have a direct contact with the products. That is why to check and verify the environmental conditions of dispensing booth is equally critical for the issuance of PPMs.

14 Production

The process of industrial-scale manufacturing of drug products in pharmaceutical industries is known as production. The overall process of drug manufacture comprises of a series of unit operations.

Different dosage forms have different series of processes. Generally the dosage forms are divided into the following main categories: Solid (tablets and capsules), Liquid (also known as oral liquid), Semi-solid (creams, ointments suppositories), and Parenteral (sterile solutions, injectable and infusion solution).

These processes are briefly described as follows.

TABLETS

Tablet manufacturing is divided into the following steps or processes: milling, granulation, mixing or blending, compression, and coating (if the finished tablet is required with coating). Some tablets do not require to be coated and therefore the coating process is not included in their manufacturing processes. The coating is often required to cover the undesired taste of the active ingredient and excipients. However, another type of coating called enteric coating is also employed in some products to protect them from getting dissolved in small intestine.

CAPSULES

Capsule manufacturing processes are a little different than the tablets as some capsules are found as granules encapsulated in the shells and therefore the manufacturing processes do not have in them further mixing or blending and obviously the product is not compressed. Instead the granules are put in the capsule shells with a machine called encapsulator. The encapsulator has specially designed dosators which facilitates the picking of a specific amount of granules to put inside one half of the capsule shell and closing it using the other half. Once the shells are closed, the product behaves almost like a tablet with the difference that the API is enclosed in the form of small granules which are dissolved inside human body almost the same manner as the tablets.

SOFT-GEL CAPSULES

The soft-gel capsule manufacturing differs from that of regular capsules so that it is a continuous method whereby two gelatin ribbons pass between twin rotating dies. As the ribbons meet, the liquid, which is prepared according to the regular procedure, to be encapsulated is precisely injected between them. The two ribbons are sealed like an oval-shaped ball.

DOI: 10.1201/9781003366393-19

The advantage of soft-gel capsules is that they are easier to swallow due to the flexible shapes, and they reduce stomach discomfort when digested. Soft-gel capsules are used to encapsulate sensitive ingredients to protect them from tampering, and are especially ideal for oils and fat-soluble formulations.

LIQUID

Liquid pharmaceutical dosage forms of preparation are simpler than tablets and capsules. In fact, in this dosage form, the main process is dissolving the active drug substance in an aqueous or non-aqueous (e.g., glycerin, ether, and alcohol) solvent. This may as well be done by suspending the drug in an appropriate medium, or by incorporating the drug substance into an oil or water phase.

In some cases, where a fatty substance like a fat-soluble vitamin is need to be dissolved in water base, a process called milling is used to finely divide the fat-soluble vitamin into tiny particles to ensure that they are mixed well and uniformly in the water solution with other active materials.

OINTMENTS AND CREAMS

Ointments and creams are also known as topical as they are mainly used on skins. A topical is basically a mixture of oil and water. Since the oil does not readily dissolve or disperse in water, so to allow this to happen, a dispersing agent called an emulsifier is added to the mixture. A topical is therefore a type of emulsion, made of a water phase and an oil phase. Once the emulsion is formed, the next step is to fill it in the recommended size of tubes.

STERILE PRODUCTS (INJECTABLE, INFUSION SOLUTION)

Sterile manufacturing is also known as "Aseptic Manufacturing and Sterile Fill-Finish".

It is a process where a drug product, containers, and closures are sterilized. Next, the sterilized products are moved to a cleanroom to combine the products. Self-contained equipment is typically used during this process. There are two types of processes for the manufacturing of injectable, also known as parenteral.

i. Terminal sterilization
ii. Aseptic filling process

Terminal sterilization provides a Sterility Assurance Level that is possible to calculate, validate, and control, and thus incorporates a safety margin. Aseptic processing is a process performed maintaining the sterility of a material that is assembled from components, each of which has been previously sterilized. However, there is no terminal sterilization involved in this process.

The aseptic process is utilized for those products where the active drug product molecule or primary packaging container is not likely to sustain the high temperature of 121 °C.

TABLETS MANUFACTURING PROCESS

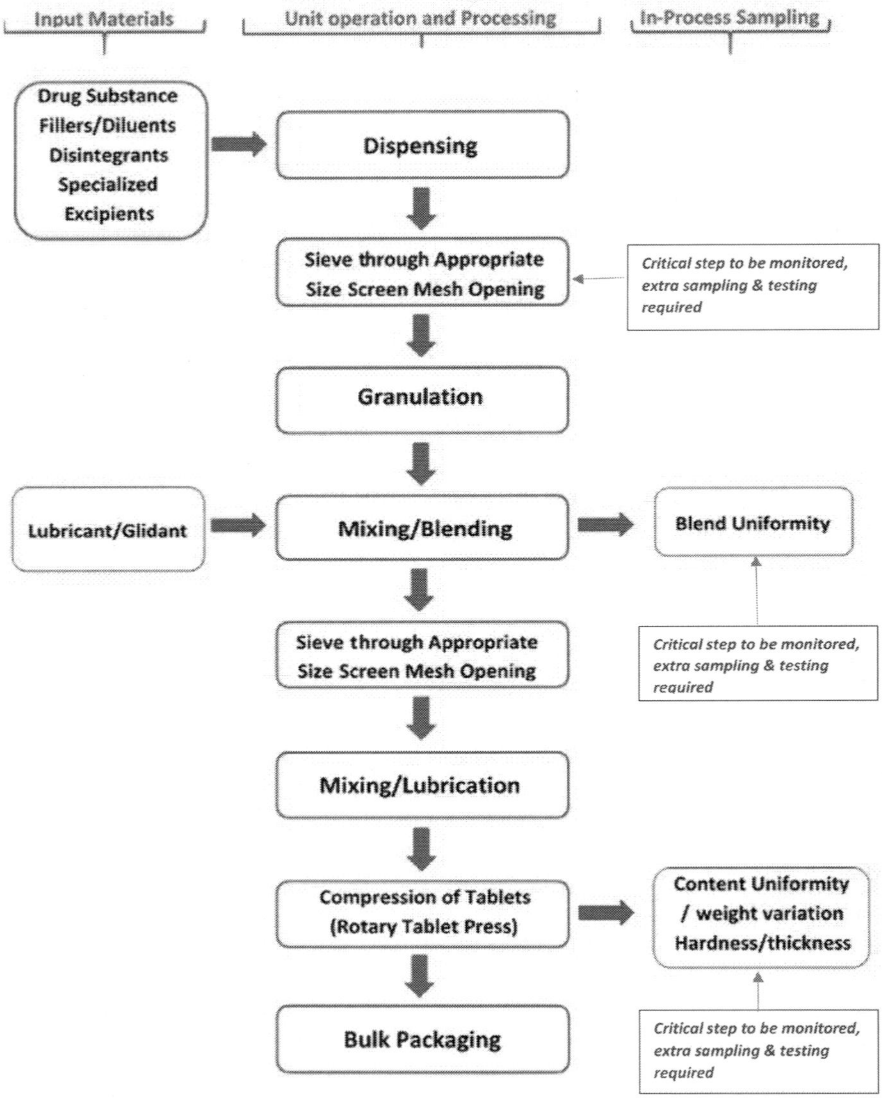

Going through the process flow diagram in Figure 1, we observe the critical steps in the manufacturing processes that need to be exclusively monitored. The best way to monitor the process is to conduct exclusive sampling and testing for a particular test. In general, the active content of the drug is tested to check and ensure that the process is in control.

LIQUID MANUFACTURING PROCESS

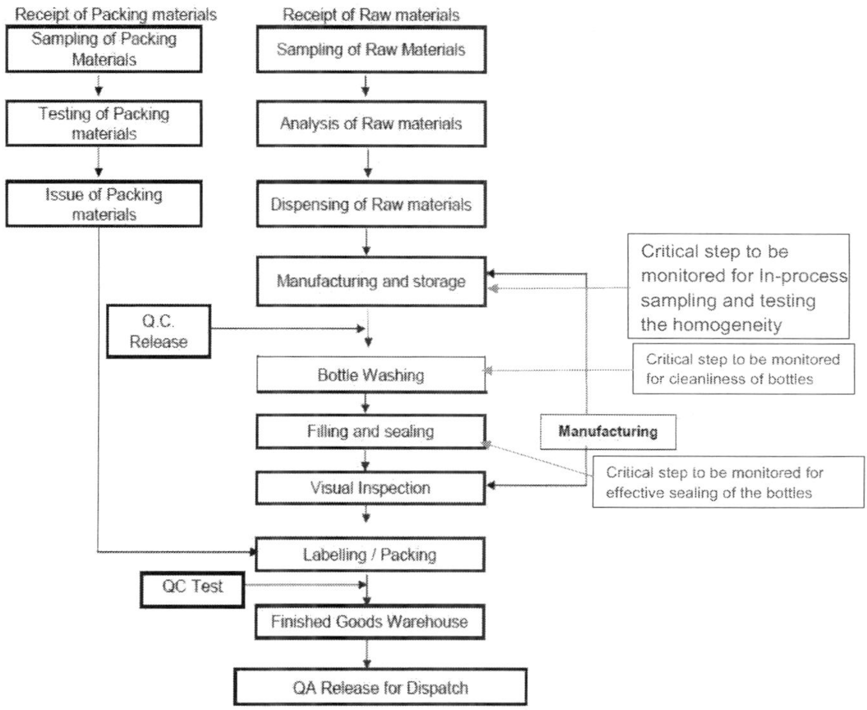

The liquid oral dosage has less critical process than tablets and capsules and the only critical step is the mixing of the active pharmaceutical ingredient that needs a surety of uniform blending (Figure 2). Depending on the nature of liquid oral, physical attributes of the solution like viscosity is also important feature that has to be controlled and monitored. However, this applies mainly in case of suspension products or multivitamins syrup where several types of water and fat-soluble vitamins are mixed together. Apart from manufacturing, filling of bottles and sealing are other important aspects that have to be under critical monitoring.

OINTMENT MANUFACTURING PROCESS

Process Flow	Control Variables	Measured Responses
Addition of raw materials Active Non actives	* Weighing / Sampling * Warehouse / QA	* Released by QC
Mixing & melting Manufacturing Vessel with Variable speed mixer Melting vessel	* Mixing time * rpm * Temperature * Final weight, yield * **Determination of Assay**	* **Smoothness & consistency** * **Assay**
Filling Automatic filling machine	* Filling machine speed * Filling tank temperature	* **Weight variation every hour**
Packaging	* packaging line speed	* Attributable characteristics critical, major, minor

Figure 3 shows the process flow diagram of an ointment manufacturing. As evident, the process is not as complicated as that of solid or liquid dosage forms. The steps and analysis shown in bold represent critical steps in an ointment or cream manufacturing.

STERILE MANUFACTURING PROCESS

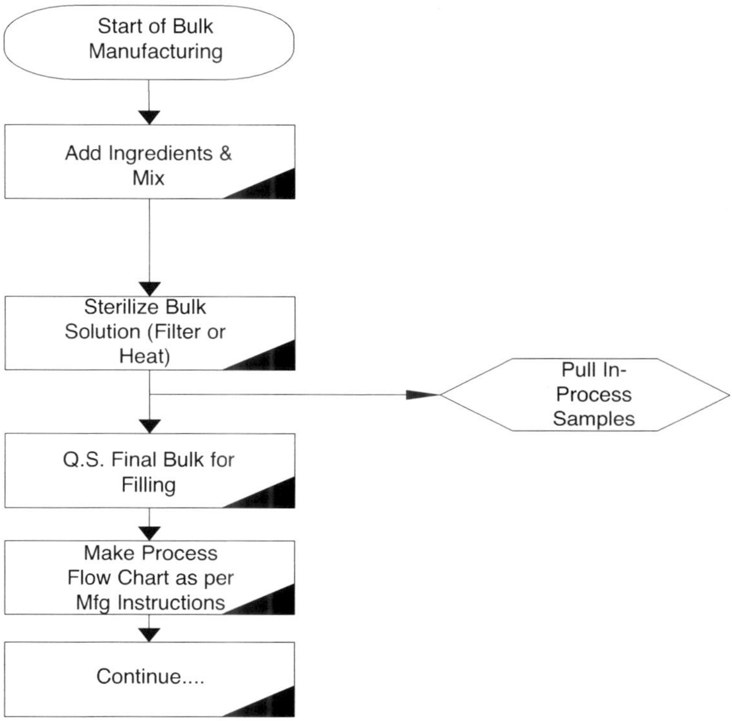

The flow diagram shown above in Figure 4 represents the manufacturing steps involved in a sterile product. The sterilization step involves either a filtration process (known as aseptic filling) through a micro filter that would not allow any viable organism or bacteria, or by heating the bulk solution by autoclaving it at 121 °C. The heated solution is later transferred to pre-sterilized containers in a separate room, classified as class A, which is free from bacteria or any viable organism. In some cases, if the containers are large (for example infusion solutions), the product is autoclaved after filling. In sterile solution manufacturing process, the only critical step where in-process sampling and monitoring are needed is often the step when the active molecule has been dissolved in water. Once the in-process sample verifies the content uniformity and solutions required pH, the filling process could be started depending upon the nature of the product if it is sterilized in bulk, filled aseptically, or terminally sterilized with containers after filling is done.

A critical aspect of sterile manufacturing area is the gowning procedure. This is important to ensure that viable and non-viable particulates do not exceed the allowed maximum limit.

Great dedication and training is required in this area as the products manufactured here are used only when the patient is in a poor physical state due to sickness or injury.

A brief overview as to how the gowning procedure is carried out is given below.

GOWNING FOR STERILE PRODUCTION

Ever wonder why specially designed gowning is important for the personnel working in the sterile or aseptic manufacturing areas?

The reason is the products that are always manufactured in cleanrooms; there are special environments that are designed and operated to remove virtually all inert particulate contaminants and viable organisms such as bacteria to save the products from being contaminated.

What are the sources of these contamination? The answer is, the greatest source of contamination is by people. Skin flakes and hair provide a home to millions of bacteria that live on us and the human.

To prevent airborne particles from contaminating the product, cleanrooms use unidirectional parallel airstreams to pick up the particles and re-cycle them through filtered air ducts. The clean air within the cleanroom is maintained at a higher pressure than externally, preventing contaminated air from re-entering. Additionally, special cleanroom garments have been designed to act as a "people" filter.

In this section, we learn how to enter the sterile production area, the background to sterile, and aseptic operations and then how to prepare and dress for working in this very special environment.

It should not be forgotten that, if the products are contaminated while working in the sterile production area, it is not just a matter of the patient returning it but it's also possible that their lives are endangered. Therefore, first the contamination and its control have to be taken care of. Manufacturing processes and appropriate gowning procedures may vary slightly at every plant but the fundamentals remain the same.

CONTAMINATION CONTROL AND CLEANROOMS

It is essential that the gowning procedure training moves on to aseptic and sterile production, and, in particular, that it is essential to follow the stringent preparation and gowning protocols necessary on every occasion before entering the aseptic suite.

This must be remembered that GMP is designed to have three outcomes, which are:

- The efficacy of the product to do what it promises
- The safety of the product not to do harm to the patients and customers
- The quality of the product, which GMP defines simply as that *the product must be completely fit for its intended use.*

All of which means that it must be:

- The product should be exactly as prescribed by the doctor
- Packaged and labeled in the correct container, also that it must be unadulterated at exactly the right strength and uncontaminated.

Obviously pharmaceutical products must not be contaminated by residues of the ingredients of other medications from equipment that wasn't cleaned properly after the last batch, brick dust from walls or floors, metal filings from worn or badly fitted machinery, pollen and, despite the extensive documentation, even the wrong ingredients. In addition, human skin and hair are two of the most common contaminators.

Particularly relevant is the fact that many of these contaminants, particularly the human sourced bits and pieces are home to bacteria and controlling viable organisms such as bacteria is the reason for sterile and aseptic working.

PREPARING AND GOWNING FOR STERILE AND ASEPTIC WORKING

- All cleanroom staff must remove their shoes and street clothing in changing rooms normally when they arrive to work, near the entrance to the building.
- Lab coats and internal shoes are worn for the move from the outside changing room to the cleanroom changing suite which are often zoned into black, gray and white areas.
- Hair coverings are worn, by all workers throughout the plant.
- While entering the changing rooms the soiling is removed from the shoe soles by the special mats at the entrance doors.
- The area where all non-permissible garments and possessions are removed and stored is known as the black area.
- The jewelry and watches are also to be removed to ensure that no loose skin flakes can be released while in the cleanroom.
- All valuables must be placed in a locker. Take care that cuffs and ankle closures of undersuits are adjusted properly, consistent with comfort.
- Now scrub-up takes place in the gray area and the cleanroom clothing is put on.
- All make-up must be removed because cosmetics are hazardous to the environment as they contain powders and greases which entice viable organisms both airborne and by transfer or cross-contamination.
- Next step is washing hands and arms up to the elbows. Do not scrub up too aggressively and keep a balance between over aggressive actions.
- Drying by warm air or sterile towels, should be thorough. Now it's time for gowning. Gloves to be put on. This is to ensure that no particles or viable organisms can cross-contaminate the outside of any of the clothing as it is donned.
- The hood comes next, over the mop cap, hair tucked in and the wrap spread carefully on the shoulders.
- This is the area where masks are put on, make sure that the ties are secure and the mask will remain in place during work.
- The critical feature when donning a coverall, especially when it has been sterilized, is to ensure that no part touches the floor during gowning.
- The hood should be worn inside the coverall, the cape of which should be smoothed inside the coverall.
- Then ensure that the front zip is done-up properly to the neck and that the wrist cuffs are adjusted as well.
- Next step is footwear, which marks the beginning of the cleanroom zone.

- For working under sterile or aseptic conditions, the over boots must be long length and knee-high.
- As each boot is put on, the leg is swung over the step-over into the white area or cleanroom zone.
- One of the rules of cleanroom behavior is that one should never touch one's clothing once dressed.
- Finally, collect the discarded clothing packaging and place it, together with your changing gloves, carefully in the waste bin.
- Put on the protective spectacles or goggles for eye protection.
- The next step is to don cleanroom working gloves. In some locations, two pairs are worn to ensure integrity should a glove become damaged.
- In the last step, apply the antibacterial disinfectant to your gloves.
- Enter the automatic airlock, a simple interlocked door with an air shower or a sophisticated air tunnel and get ready to work in the cleanroom.

LEAVING THE CLEANROOM

The procedure for leaving the cleanroom, either for breaks or at the end of the shift, is basically a mirror image of entering but in reverse order. The route out should be different that the workers must leave through a dedicated uncovering area and place their cleanroom clothing in appropriate containers for decontamination processing.

If there is no provision for change into different clothing on each entry than the route will take the employees through a specific area where their cleanroom clothing will be stored and will be hung-up ready for their next entry.

After this, change out of the undersuits again, placing them in a laundry container if changed every time or in the storage area if they are re-usable.

Finally, in the cleanroom suite, change into your domestic clothing and internal uniform and recover your outdoors clothing at the outdoors changing room.

QUESTIONS AND ANSWERS

1. **What is the basic difference in regular capsules and soft-gel capsules manufacturing processes?**
   Capsules are dosage forms in which granules are encapsulated in the shells using a special kind of machine called "encapsulator". Whereas the soft-gel capsules manufacturing process differs from that of regular capsules so that it is a continuous method whereby two gelatin ribbons pass between twin rotating dies. As the ribbons meet, the liquid, which is prepared according to the regular procedure, to be encapsulated is precisely injected between them. The two ribbons are then sealed like an oval-shaped ball.

2. **What are the advantages of using soft-gel capsules over regular capsules?**
   Answer: The flexible shapes of soft-gel capsules make them easier to swallow. They cause less discomfort to stomach when digested. Also, sensitive ingredients are used to encapsulate in soft gels to protected from tampering. Softgels are ideal for oils and fat-soluble formulations.

3. What is the name of the material used to facilitate the mixing oil and water in the manufacture of topicals?

Answer: A topical is basically a mixture of oil and water. Since the oil does not readily dissolve or disperse in water, so to allow this to happen, a dispersing agent called an emulsifier is added to the mixture. A topical is therefore is a type of emulsion, made of a water phase and an oil phase.

4. Name the various processes involved in the manufacture of a tablet product?

Answer: Tablet manufacturing comprises of the following steps;

Milling, granulation, mixing or blending, compression, and coating (if the finished tablet is required with coating).

Some tablets do not require to be coated and therefore the coating process is not included in their manufacturing processes.

5. What is the name of the process used to mix fat-soluble materials in water base to manufacture a liquid product?

Answer: Since the liquid is prepared mainly by dissolving the active drug substance in an aqueous or non-aqueous (e.g., glycerin, ether, alcohol) solvent, it sometimes becomes difficult when a fatty substance like a fat-soluble vitamin needs to be dissolved in the solution. This could be done by suspending the drug in an appropriate medium, or by incorporating the drug substance into an oil or water phase. Otherwise, a fat-soluble vitamin is made to go through a process called milling to finely divide the fat-soluble vitamin into tiny particles to ensure they are mixed well and uniformly in the water solution with other active materials.

6. What are the different ways to prepare a sterile product?

Answer: There are two types of processes for the manufacturing of sterile products or injectable, also known as parenteral.

i. Terminal sterilization
ii. Aseptic filling process

Terminal sterilization is achieved by exposure to temperature for a predetermined extent of treatment.

Aseptic process is utilized for those products where the active drug product molecule or primary packaging container is not likely to sustain the high temperature of 121 °C.

Therefore, aseptic processing is performed maintaining the sterility of a material that is assembled from components, each of which has been previously sterilized. However, there is no terminal sterilization involved in this process.

7. What is the meaning of critical steps in manufacturing processes and why is it important to control them?

Answer: A critical step in the manufacturing process of a dosage form is the process which have an impact on a critical quality attribute of the product and which, if not controlled, may result in compromised quality, safety and efficacy of the end product.

8. **Why is gowning necessary to enter a sterile area?**
   Answer: The reason for gowning is that the injectable products are manufactured in special environments that are designed and operated to remove virtually all inert particulate contaminants and viable organisms such as bacteria to save the products from being contaminated. Since human beings are considered as the greatest source of contamination, it's important to cover the entire body using special gowning techniques and stop the skin flakes and hair, which are homes to millions of bacteria, from falling in the cleanrooms and contaminating the area.

9. **What would happen if contamination is not prevented in the cleanrooms?**
   Answer: If the products are contaminated while working in the sterile production area, it is not just a matter of the patient returning it but it's also possible that their lives are endangered. Therefore, the contamination and its control have to be strictly taken care of.

10. **What are the three main outcomes of GMP?**
    Answer: GMP is designed to have three outcomes, which are;

    - The efficacy of the product to do what it promises
    - The safety of the product not to do harm to the patients and customers
    - The quality of the product, which GMP defines simply as that *the product must be completely fit for its intended use.*

15 Packaging

What is the main objective of pharmaceutical packaging? Of course not for just showing colors and designs to enhance the sale of the products but certainly to store and transport pharmaceutical products safely. It is rather to protect the product from physical, chemical, and microbial invasion. It must maintain the physical quality of all dosage and protect against all damage and breakage.

Pharmaceutical packaging refers to the operation of filling, sealing, packaging, and labeling semi-finished products or finished pharmaceutical products with appropriate containers, and using packaging technology to provide quality assurance, identification of trademarks, and descriptions for drugs. The role of drug packaging in the field of circulation can be divided into two categories: inner packaging and outer packaging. Its functions mainly have three aspects;

- Protection function
- Convenient application, and
- Product promotion.

Packaging maintain the characteristic properties of the pharmaceutical products.

Pharmaceutical packaging refers to packaging materials and containers that belongs to the category of special packaging. It must have all the attributes of packaging and has its particularities as mentioned below:

1. Protection of the drug from environmental effects during storage and use, and maintain the original properties of the drug.
2. Drug packaging materials themselves should have certain stability during storage and use.
3. Drug packaging materials must not pollute the drug production environment when wrapping drugs.
4. The packaging materials should be inert and must not react chemically or biologically with the packaged drugs.
5. The pharmaceutical packaging materials must be tested and released by quality control to ensure they can be used to package drugs.

For that matter, pharmaceutical company first needs to think about what packaging is required in order to deliver a product successfully and safely. This largely depends on what product we are dealing with.

In order to understand the concepts and ideas behind the packaging process lets first understand the types of packaging that pharmaceutical products are packed in.

DOI: 10.1201/9781003366393-20

There are three stages of packaging in pharmaceutical industry:

- Primary packaging
- Secondary packaging and
- Tertiary packaging.

There are some medical products that require only primary packing. However, there are many that need all three of primary, secondary, and tertiary packaging. These are the ones which cannot afford to carry a risk that may lead to any kind of alteration in the chemical composition of the dosage. Also, there might be some risks during the transport of the products; therefore, tertiary packaging helps to protect them during storage and transport. It's extremely essential to protect pharmaceutical products from damage during shipping and distribution, as faulty medication could have disastrous consequences for consumers and can even cause a market return leading to a huge financial loss to the company.

All three types of packaging are important and have their own significance in pharmaceutical packaging line. That's why the drug packaging process is so important, and must be carried out correctly.

PRIMARY PACKAGING

Primary packing are the ones where the packaging materials come in direct contact with pharmaceutical products. Examples of primary packaging are blister packs, strip packs, bottles, vials, and ampoules. Primary Packaging are the main line of defense between a drug and its environment, therefore, primary packaging must be able to prevent contamination and chemical alterations. It is usually made of materials that are impermeable to air, light, and moisture to protect the product from the environment. In the pharmaceutical industry, primary packaging must meet strict regulations to ensure the safety and quality of the product.

VIALS AND AMPOULES

Of the primary packaging material, the sterile products are stored in ampoules and vials which are small sealed glass containers that are often used to contain a single dose of a medication and are capable of being terminally sterilized at 121 °C. Ampoules have a pointed end that is broken off when the ampoule is opened. In addition to glass, vials are also made of plastic and have a rubber stopper that is punctured when the vial is opened.

BLISTERS PACKS

The solid dosage forms like tablets, capsules, and suppositories are commonly packed in blisters as primary packaging. Although there are glass and plastic bottles also used to pack tablets and capsules, many companies still prefer blister packs.

Blister packs are pre-formed plastic/paper/foil packaging used for formed solid drugs. The primary component of a blister pack is a cavity or pocket made from a thermo-formed plastic. This usually has a backing of paperboard or a lidding seal of aluminum foil or plastic film. Blister packs are useful for protecting drugs against external factors, such as humidity and contamination for extended periods of time.

STRIP PACKS

The main difference between blister and strip packaging is that strip doesn't have thermo-formed cavities. Instead, the packaging is formed around the tablet. The purpose of strip packs is that it is specifically used to increase the dosage life as it protects the content individually.

BOTTLES

Liquid non-sterile products are commonly packed in glass bottles. Glass is widely used to carry liquid products because of its inert nature and other excellent properties.

Although they come in different colors, the most common is brown and orange as they can prevent ultraviolet light from harming photosensitive contents inside.

SECONDARY PACKAGING

Once a drug has undergone primary packaging, secondary packaging can begin. It is usually made of materials that are less durable than primary packaging, such as cardboard or paperboard.

Secondary packaging is the second layer of packaging that protects the product from damage during shipping and handling.

One of the main purposes of secondary packaging is to provide printing space for both company branding and product information. Secondary packaging plays an important role in making products consumer friendly, as it is typically used to display useful information like the API name, strength of unit dose, manufacturer's name, manufacturing and expiry dates, directions, warnings, etc.

TERTIARY PACKAGING

This is the third layer of packaging and in most cases the last one. It is mainly used for the shipping process. The main objective of tertiary packaging is safeguarding primary and secondary packaging from the external environment during storage and transportation. The most commonly used tertiary packagings are plane boxes, cardboards, and shrink wraps.

The end consumers don't get the products in this packaging as this packaging is removed by the retailers before the product is put in the market or pharmacy for commercial purpose.

MOST EFFECTIVE PHARMACEUTICAL PACKAGING MATERIALS

Pharmaceutical packaging materials refer to the materials used in the manufacture of packaging containers, packaging, and decoration, packaging and printing, packaging and transportation, etc. to meet product packaging requirements. There are different types of packaging materials used in the healthcare industry; however, some materials achieve better results than others. These materials help ensure that pharma drugs maintain optimal conditions during transportation and arrive safely to the end user. Some of the most popular materials for packaging options include:

- plastic films
- aluminum foils
- glass
- plastic and elastomer caps
- tubes made of aluminum
- plastic and composite materials
- containers made of plastic

Aluminum foil is used for blister packaging of tablets and capsules, and for the packaging of some aerosol products. The foil is an impenetrable barrier to oxygen, moisture, light, and microorganisms. Due to its superior barrier properties, aluminum films are considered a best choice for pharma packaging.

Plastic and polymer-based materials include pharmaceutical sachets, polybags, pre-fillable syringes, and blister packs that are used for tablets, capsules, powders, and granulates. These materials are not only durable but also cost-effective; however, the materials may interact with certain pharmaceutical drugs.

Glass containers include vials, pre-filled syringes, and bottles which are used for dry powder and liquid-based medicines. Their transparency, easy visual inspection, and protection against contaminants have made them ideal for pharmaceutical packaging. Additionally, they are easy to sterilize and recycle.

QUESTIONS AND ANSWERS

1. **What is pharmaceutical packaging?**
   Answer: Pharmaceutical packaging refers to the operation of filling, sealing, packaging, and labeling semi-finished products or finished pharmaceutical products with appropriate containers, and using packaging technology to provide quality assurance, identification of trademarks, and descriptions for drugs.

2. **What are the advantages of using glass containers for pharmaceutical packaging?**
   Answer: Glass containers which include vials, pre-filled syringes, and bottles are used for dry powder and liquid-based medicines. The advantages of using glass containers are that they are transparent, easy to carry out visual inspection, and provide effective protection against contaminants. Additionally, they are easy to sterilize and recycle.

3. **What is primary packaging of pharmaceutical drugs and why is it critical?**
   Answer: Primary packaging is the main line of defense between a drug and its environment. It is critical because the packaging materials come in direct contact with pharmaceutical products. Therefore, primary packaging must be able to prevent contamination and chemical alterations. It is usually made of materials that are impermeable to air, light, and moisture to protect the product from the environment. In the pharmaceutical industry, primary packaging must meet strict regulations to ensure the safety and quality of the product.

4. **Name the most effective packaging materials?**
   Answer: There are different types of packaging materials used in the healthcare industry; however some materials achieve better results than others. These materials help ensure that pharma drugs maintain optimal conditions during transportation and arrive safely to the end user. Some of the most popular materials for packaging options include the following:

   - plastic films
   - aluminum foils
   - glass
   - plastic and elastomer caps
   - tubes made of aluminum
   - plastic and composite materials
   - containers made of plastic

5. **What are some specific attributes of the packaging materials?**
   Answer: The packaging materials in pharmaceutical industry must have the following attributes as mentioned below:

   1. Protection of the drug from environmental effects during storage and use, and maintain the original properties of the drug.
   2. Drug packaging materials themselves should have certain stability during storage and use.
   3. Drug packaging materials must not pollute the drug production environment when wrapping drugs.
   4. The packaging materials should be inert and must not react chemically or biologically with the packaged drugs.
   5. The pharmaceutical packaging materials must be tested and released by quality control to ensure that they can be used to package drugs.

6. **What is the difference between a blister pack and a strip packaging?**
   Answer: The main difference between blister and strip packaging is that strip doesn't have thermo-formed cavities. Instead, the packaging is formed around the tablet. The purpose of strip packs is that it is specifically used to increase the dosage life as it protects the content individually.

16 Role of Quality Assurance

Pharmaceutical Industries continued to act with a reactive approach of quality control for long. The purpose of quality control was, and still is, to ensure that the finished product going to the market for commercial purpose is tested to comply before release. However, there was no control over the entire series of processes starting from dispensing to packaging which would prevent a machine, material, or human mistake or error. Therefore, the purpose of quality control was limited to the function of analyzing the finished product and ensure that they were manufactured correctly. In case of a non-compliance result, the entire batch could be either scrapped or if possible reworked and salvaged to make them compliant.

The introduction of quality assurance brought a revolutionary change in the pharmaceutical industries in terms of regulatory compliance. It was identified as an approach that creates the environment by ensuring compliance of machine, material, and man before they get involved in the process of manufacturing a medicinal product. Qualification and validation were later introduced under the umbrella of quality assurance to make the regulatory requirements and standards more stringent for pharmaceutical industries.

As a bottom line, the role of quality assurance is to ensure that the medication being manufactured will provide the desired effect to the patient and that the medications will meet quality requirements and all relevant regulations.

In the pages to follow, we will see the role and responsibilities of quality assurance in more detail.

IMPORTANCE OF QUALITY ASSURANCE

Why quality assurance is important? To answer this question, we have to understand the total responsibilities of QA. The department of QA analyzes and verifies the quality of a product throughout the design, development, and production process. As a matter of fact, the responsibility starts from the moment a raw material or Active Pharmaceutical Ingredient arrives in the company. There are several different ways the QA associates utilize the principles and methodologies to improve the overall consistency of the manufacturing operations at every stage. Implementing and improving quality assurance practices can help improve a company's reputation, increase customer confidence, and increase efficiency.

By successfully implementing Quality assurance and its standards, companies can identify any potential problems with the product. It can isolate an element of the manufacturing process which can be quickly fixed. It can also play a large role in controlling costs as it highlights processes that are inefficient or expensive, identifies,

DOI: 10.1201/9781003366393-21

and rectifies mistakes and errors during the processes, ultimately avoiding re-works, and salvaging or discarding the batch leading to saving a huge amount of time, money, and material.

DEFINITION OF QUALITY ASSURANCE

There are various ways to define the quality assurance; however, a simple definition by International Organization for Standardization is that "quality assurance as a part of quality management focused on providing confidence that products fulfill agreed quality requirements".

DIFFERENCE BETWEEN QC AND QA

The role of Quality assurance is different from quality control as instead of just at the end like quality control, it happens throughout the entire product lifecycle. It looks at every step in the development process. Whatever stage of the process, companies have an obligation to make sure that the production of their goods is safe and ethical.

QA'S ROLE IN ENSURING PRODUCT'S SAFETY

If a product is defective or hazardous, it could have major consequences for the consumer and the business. Since the principles of QA are crucial in pharmaceutical industries, the QA associates can test the safety and efficiency of products in several ways throughout the production process. This even includes verification of equipment maintenance conditions and their operational and functional state by ensuring the necessary calibration and adherence to preventative maintenance program. Besides the equipment status, the validity of materials used, quantity dispensed, environmental conditions check to verify that no pollutants or cross contamination enter the production space, personnel trainings and experiences are a few other critical aspects to be checked and verified by the QA department. Alongside this, the patients and customers also benefit from knowing that the products they purchase meet all statutory or regulatory requirements. Consistent levels of quality are crucial for a business, and quality assurance is the process that ensures this.

In order to ensure products safety and quality, QA has to perform the monitoring and controlling of all processes, conducting quality audits and QA reviews, documenting and approving the standard operating procedures and hence providing management and control to the processes.

QA also allows team leaders to identify any faulty elements in the production process and address them. It's important to have a schedule for QA checks, developed by team members with broad experience in the supply chain.

QUALITY ASSURANCE STANDARDS

It's important to understand that products have to meet the quality assurance standards comprising of both international and industry standards.

The International Organization for Standardization (ISO) has created specific guidelines for products and safety requirements to ensure the safety and efficiency of the products. Besides the ISO standards other international regulatory requirements are set by the World Health Organization (WHO), United States FDA, and European Union (EU) regulations. An effective QA department of a pharmaceutical company has to establish its own internal standards that may encompass or reflect the requirements of all the international standards. QA also facilitates certification and approval from international organizations and Health regulators and maintain them in order to enhance the image of the company in producing quality and safe products for customers.

METHODOLOGIES OF QUALITY ASSURANCE TESTING

The most common of QA testing methodologies are as follows:

1. **Ishikawa Diagrams (also known as fish bone diagram):** used as a cause and effect diagram to analyze possible causes of a problem and in sorting ideas into useful categories.
2. **Benchmarking:** QA uses benchmark, a set of standards to evaluate the quality of a product or make gradual improvements to it
3. **Standard Operating Procedures:** Documents often initiated by and totally controlled by QA, which are specific to related operations and describes the activities necessary to complete tasks in accordance with industry regulations
4. **Statistical Sampling:** Statistical sampling is a critical method for testing in QA. It involves drawing a set of observations randomly from a population distribution.
5. Process reviews:

QUESTIONS AND ANSWERS

1. **Why quality assurance is different from quality control?**
   Answer: The purpose of quality control was limited to the function of analyzing the finished product and ensure that they were manufactured correctly. In case of a non-compliance result, the entire batch could be either scrapped or if possible reworked and salvaged to make them compliant. However, the introduction of quality assurance brought a change in the in terms of regulatory compliance. It was identified as an approach that creates the environment by ensuring compliance of machine, material and man before they get involved in the process of manufacturing a medicinal product.

2. **Define quality assurance:**
   The simple definition of quality assurance by International Organization for Standardization is that "quality assurance as a part of quality management focused on providing confidence that products fulfill agreed quality requirements".

3. **What are the various ways that QA uses to ensure product's safety and quality?**
   Answer: The principles of QA are crucial in pharmaceutical industries, the QA associates can test the safety and efficiency of products in several ways throughout the production process. This includes but not limited to verification of equipment maintenance conditions and their operational and functional state by ensuring the necessary calibration and adherence to preventative maintenance program. Besides the equipment status, the validity of materials used, quantity dispensed, environmental conditions to verify that no pollutants or cross contamination enter the production area, personnel trainings and experiences, are other critical aspects to be considered and verified by the QA department to ensure product's safety and quality.

4. **What is a fish bone diagram used for?**
   Answer: Fish bone diagram, also called as "Ishikawa diagrams", is one of the various methodologies for QA used as a cause and effect diagram to analyze possible causes of a problem and in sorting ideas into useful categories.

5. **What is statistical sampling?**
   Answer: Statistical sampling is a critical method for testing in QA. It involves drawing a set of observations randomly from a population distribution.

17 Role of Quality Control

The quality department is vital in the pharmaceutical industry to ensure that pharmaceutical products are safe, effective, and of high quality. It helps protect public health, maintain regulatory compliance, and build trust in the industry.

Quality control and quality assurance are often used interchangeably in pharmaceutical settings. This is because they both refer to aspects of quality management and meeting customers' high expectations and demands.

Without these two quality management functions, a pharmaceutical organization would struggle to achieve consistency in its output. Quality control and quality assurance are highly important and contribute to productivity and the overall success of an organization's products.

QUALITY CONTROL (QC)

Quality Control is that part of Good Manufacturing Practice where raw materials, packaging components, in-process samples, and finished products are tested. Based on the specification, this department determines if a product is failing or passing.

The QC laboratory tests samples of raw materials and finished products. It is critical to get accurate results from QC tests. Otherwise, defective batches may be released, and good batches may be rejected.

The role of the QC laboratory is to detect any possible defects once they have occurred rather than to prevent them from occurring.

Therefore, quality control is the most important part of the pharmaceutical industry.

QUALITY ASSURANCE (QA)

From the time when raw materials are received in a company up to the release of the finished product, quality assurance ensures that all the processes meet the specifications and that GMP is followed throughout the process.

ROLE OF QUALITY CONTROL

Quality control is critical in the pharmaceutical industry to ensure pharmaceutical products' safety, efficacy, and consistency. It involves various activities and processes to monitor and control the quality of packaging components, raw materials, intermediate products, and finished pharmaceutical products.

The safety, quality, efficacy, and benefit of compounded preparations depend on correct ingredients and calculations; accurate and precise measurements; appropriate formulation, facilities, equipment, and procedures; and prudent pharmaceutical judgment.

DOI: 10.1201/9781003366393-22

Quality control is responsible for arranging sound and appropriate specifications, standards, sampling plans, and test procedures designed to assure that components, drug product containers, closures, in-process materials, labelling, and drug products conform to appropriate standards of identity, strength, quality, and purity.

Preparation and Implementation of Standard Operating Procedures (SOP)

SOPs for pharmaceutical testing describe how to perform routine and expected tasks in the laboratory environment.

SOPs provide clear instructions that describe when a task will be performed, how a task will be performed, who will perform the task, why the task is necessary, any limitations in performing the task, and what action to take when unacceptable deviations or discrepancies occur.

SOPs must be reviewed regularly and updated as necessary. Auditing and verifying compliance with established SOPs should be performed periodically. The SOP should be specific to each device and process used in testing. Properly maintained and implemented SOPs are vital to preparation quality.

Compliance with Regulations

The pharmaceutical industry is highly regulated. Quality control ensures that pharmaceutical companies comply with various regulatory requirements and guidelines, such as those set by the Food and Drug Administration (FDA) in the United States or the European Medicines Agency (EMA) in Europe. It implements and maintains quality systems that meet the current Good Manufacturing Practices (cGMP) standards.

Test Methods and Specifications

Management has to provide specific validated test methods for each test and product.

Test methods provide specific directions on properly executing a test procedure in a standardized manner. A test method is specific to analysis and instrumental techniques such as HPLC, UV, or GC. Specifications generally accompany test methods and provide criteria to pass or fail a particular test and acceptance criteria for a test method, such as system suitability or control limits.

Laboratory Testing Records

Sample and reagent preparation should be documented in laboratory notebooks. Notebooks document the preparations of laboratory solutions, standards, and working reagents. It is important in a laboratory to provide accurate instructions and records of these preparations.

Testing records include laboratory analyst notebooks, specific testing sheets, analytical printouts, electronic records such as chromatographs, and ancillary records that support the laboratory's compliance. Records would include calibration reports, training records, and monitoring of the environment.

Training

A quality control program for testing medicines requires training documentation of both training and skill competency.

Personnel involved in nonsterile or sterile testing requires additional, specific training, and periodic retraining beyond the training needed for routine duties.

If testing is outsourced, the management must be assured of the credentials, proper training, and continuing competency activities of the personnel in the contract laboratory.

Documentation

The purpose of documentation is to provide a record of all aspects of testing operations and procedures.

Information on the testing record should ideally be entered as the tasks are performed or as testing data is received. testing records should be reviewed for accuracy and completeness (as appropriate) and approved by QA personnel.

Verification

Verification involves authoritatively signed assurance and documentation that a process, procedure, or equipment is functioning properly and producing the expected results. Verifying a testing procedure involves checking to ensure the calculations; weighing and measuring, order of mixing, and testing techniques and equipment are appropriate and accurately performed.

The quality of ingredients should be verified upon receipt. For example, Certificate of Analysis, manufacturer's label on commercial products, etc.

Equipment verification methods are sometimes available from specific equipment manufacturers or can be developed in-house. The responsibility for verifying equipment performance, including work completed by contractors.

Testing of Raw Materials

Quality control involves testing raw materials, packaging components, and in-process and finished products used in pharmaceutical production. This includes testing the identity, purity, potency, and quality of active pharmaceutical ingredients (APIs), excipients, and other materials.

In-Process Control

Quality control monitors the whole manufacturing process to ensure it is performed correctly and meets specifications. In-process control involves testing and verifying critical parameters, such as temperature, pH, and moisture content, at different stages of production. This helps detect and prevent any deviations from the expected quality, ensuring the consistency and quality of the final product.

Finished Product Testing

Quality control conducts testing of finished pharmaceutical products to assess their quality, safety, and efficacy. This includes various analytical tests, such as assay, impurities, dissolution, stability, and microbial testing, to ensure the product meets the predetermined standards and specifications. Finished product testing helps identify deviations or abnormalities affecting the product's quality and performance.

Data Interpretation

The collection of raw data from the testing process must be completed accurately. The analyst must ensure that appropriate and valid descriptive statistics are used to analyze the data and that the operating parameters of the analytical instruments are well-established. Reference values, if available, should be provided with the analytical results. A description of the analytical controls used by the laboratory is important for documentation, as is the source of reference standards used to establish standard curves.

Documentation and Recordkeeping

Quality control maintains detailed documentation and records of all testing activities, results, and procedures. This documentation is a critical reference for verifying compliance with regulatory requirements and providing traceability throughout the manufacturing process. It allows for a thorough investigation of quality issues or recalls and provides evidence of product quality for regulatory audits.

Corrective Action and Preventive Action (CAPA)

The pharmaceutical company should have a system for implementing corrective and preventive actions resulting from investigating complaints, product rejections, non-conformances, recalls, deviations, audits, regulatory inspections and findings, and trends from process performance and product quality monitoring…

Continuous Improvement

Quality control plays a crucial role in continuous improvement efforts within the pharmaceutical industry. It monitors quality metrics, conducts trend analysis, and identifies areas for improvement. By implementing corrective and preventive actions based on quality control data, pharmaceutical companies can enhance their manufacturing processes, reduce deviations, and enhance overall product quality.

Change Control System

The change control system ensures that continual improvement is undertaken promptly and effectively. It should provide a high degree of assurance that the change has no unintended consequences.

Storage Requirements

Storage requirements for samples and standards must be specified, including the type of container, temperature, humidity, and light protection.

Before handling the sample, the effect(s) of any substance on the sample preparation that may interfere with or alter the results must be known. When sending a preparation to a contract laboratory, they should provide the complete written formulation so that the laboratory can quickly determine if any interfering substances are present.

Microbiological Testing

Microbiological testing for pharmacy compounding includes sterility, endotoxin, preservative effectiveness, and microbial limit testing.

Sterility Testing

Sterility tests may be conducted using commercial kits or by developing and verifying USP sterility testing protocols.

Quality Risk Management

Quality risk management is a systematic process for the assessment, control, communication, and review of risks to the quality of a drug across the product lifecycle. It can be applied both proactively and retrospectively.

According to the "International Council for Harmonisation of Technical Requirements for Pharmaceuticals for Human Use (ICH)", quality risk management has two primary principles,

- Quality risk evaluation should be based on scientific knowledge, such as data and research, and ultimately work towards the protection of consumers.
- The higher the risk level, the stronger a QRM process should be. The level of effort, formality, and documentation should be equal to the risk level.

These principles can be seen in the examples provided below by the European Commission in its Good Distribution Practice (GDP) Guidelines for Medicinal Products:

- Temperature mapping should be repeated according to the risk assessment results.
- Equipment should be calibrated at defined intervals based on a risk and reliability assessment.
- The scope and extent of key equipment qualification and/or key process validation must be determined using a documented risk assessment approach.
- All means available must be used to minimize the risk of falsified medicinal products entering the supply chain.

- Returned products must be handled according to a documented, risk-based process.
- Risk assessment of delivery routes should be used to determine where temperature controls are required.

WORK FLOW OF QC TESTING

Sample Receiving

The sample received in QC should be login. There should be a document /register where sample identification, receiving date and time should be documented. This should be a traceable record, and the sample could be tracked.

Work Assignment

The sample should be assigned to the analyst(s) for testing. The method, specification, and other resources should be provided to the analyst.

Samples/Standard Preparation and Instruments. Set Up

The test method should describe how to prepare the sample and standard. This may involve sample dilution or extraction. A reference to the current test method should be documented before setting up the sample.

Testing the Sample

Once the sample and standard are prepared, it is included in the test run. The sample should be placed under controlled conditions to prevent any degradation. The sample identification must be documented in instrument logbooks. All preparations must be documented in the lab book as it was done.

Verification of System Suitability

Instruments must be calibrated. The methods may have some system suitability requirements. All those requirements must meet as per specification. The test methods should describe how the integrity of particular test runs is verified. Non-instrumental methods may also include control or equivalent verification systems.

Calculation and Results

After running samples and standards on an instrument, the analyst should calculate the final results. Calculations should be double-checked, and all entries in the instrument if the instrument does the calculation.

Out of Specifications (OOS)

There should be a written procedure if OOS results are produced. All OOS or atypical results must be informed to management to conduct the necessary investigation and inform the concerned department to take appropriate action.

Report Result

Once the sample has been run, either the instrument or the analyst will perform a calculation to arrive at a result. It is essential to double-check all calculations before finalizing sample results.

OVERALL SUMMARY

Quality is everyone's job. Quality is important for every aspect of pharmaceutical manufacture and packaging. Quality should not be tested into products; it must be built in at each manufacturing step.

Quality must be followed during actual processing steps and in other areas such as vendor assurance, personnel training, internal audits, change management, release for supply, and QC sampling.

Pharmaceutical manufacturers must integrate the key functions of quality control and regulations or GMP under quality assurance. QA is charged with preventing defects and errors by overseeing all aspects of producing a quality product.

Overall, quality control is vital in the pharmaceutical industry to ensure that pharmaceutical products are safe, effective, and high quality. It helps protect public health, maintain regulatory compliance, and build trust in the industry.

QUESTIONS AND ANSWERS

1. **Why Quality role is important?**
   Answer: The quality department is vital in the pharmaceutical industry to ensure that pharmaceutical products are safe, effective, and of high quality. It helps protect public health, maintain regulatory compliance, and build trust in the industry.

2. **What is the basic responsibility of Quality Control?**
   Answer: It involves various activities and processes to monitor and control the quality of packaging components, raw materials, intermediate products, and finished pharmaceutical products.

3. **What is an SOP?**
   Answer: SOPs are clear instructions that describe when a task will be performed, how a task will be performed, who will perform the task, why the task is necessary, any limitations in performing the task, and what action to take when unacceptable deviations or discrepancies occur.

4. **Who provides the test method? What does a test method describe?**
Answer: Management has to provide specific validated test methods for each test and product.
Test methods provide specific directions on properly executing a test procedure in a standardized manner. A test method is specific to analysis and instrumental techniques such as HPLC, UV or GC.

5. **If OOS results are produced, what are the responsibilities of analysts and management?**
Answer: If OOS results are produced the analyst will inform management. Management will conduct the necessary investigation and inform the concerned department to take appropriate action.

6. **Why are finished products tested in QC?**
Answer: Quality control conducts testing of finished pharmaceutical products to assess their quality, safety, and efficacy. This includes various analytical tests, such as assay, impurities, dissolution, stability, and microbial testing, to ensure that the product meets the predetermined standards and specifications. Finished product testing helps identify deviations or abnormalities affecting the product's quality and performance.

7. **Why in-process controls are important? What are the most common tests being done during the in-process?**
Answer: In-process control involves testing and verifying critical parameters, such as temperature, pH, and moisture content, at different stages of production. This helps detect and prevent any deviations from the expected quality, ensuring the consistency and quality of the final product.

8. **What is quality risk management?**
Answer: Quality risk management is a systematic process for the assessment, control, communication, and review of risks to the quality of a drug across the product lifecycle. It can be applied both proactively and retrospectively.

9. **What are the two principles of risk management of ICH?**
Answer: (1) Quality risk evaluation should be based on scientific knowledge, such as data and research, and ultimately work towards the protection of consumers. (2) The higher the risk level, the stronger a QRM process should be. The effort, formality, and documentation level should equal the risk level.

18 Logistics – Cold Chain Supply

Maintaining the quality of a pharmaceutical product is not just limited up to the stage of manufacturing. It should never be considered that once the goods are out of the premises of the manufacturing company, the responsibilities of the manufacturer to ensure a quality product to reach the customer or patient which maintains its efficacy up to the end of the shelf life, are over. Because it is not. The manufacturer has to make sure that during the transportation of the product, depending on the dosage forms, different forms of liabilities still lie on their shoulders and it is their job to ensure a safe and smooth transition of the products from their stores to the distribution warehouse and from there when it goes to the market where they can be picked up by a customer at some point of time, the quality of the product is never compromised.

The quality of the product, especially a liquid or a sterile medicine, can be compromised in different ways. It's not just the potency of the product but also the packaging that matters when it reaches to the customer. To maintain the pharmaceutical products' integrity and safety, the products must stay within the recommended ranges of temperature and humidity.

The smallest temperature or humidity variation can alter the product's quality which can put the customer's health at risk and cost the pharmaceutical company a significant amount of money. During the supply chain process, there are many factors to consider in order to ensure the drugs and vaccines remain safe and follow shipping regulations from both the manufacturer and the health regulatory bodies.

In order to keep pharmaceutical products safe during transport, companies have to optimize their supply chain by ensuring that their goods are constantly stored under suitable conditions. To meet these requirements, companies that transport pharmaceuticals use a process known as cold chain shipping, or cold chain pharma logistics. Managing the pharmaceutical supply chain is complex with greater demands for business innovation, security, reducing costs, and improving customer satisfaction. Therefore, pharmaceutical companies have been facing increased challenges presented by strict regulations, costly recalls, counterfeiting, and product protection across the supply chain.

COLD CHAIN SUPPLY

The cold chain for pharmaceuticals requires uninterrupted temperature control on the journey from their site of manufacture all the way to where they are administered. As a consequence, dedicated solutions are required for the freezing/thawing, packaging, tracking, and storing of these temperature-sensitive products. Therefore, cold chain logistics are essential to transport the pharmaceutical and bio

DOI: 10.1201/9781003366393-23

pharmaceutical products. High temperature may cause the degradation in these products which could harm the end user. Over the years the pharmaceutical industry has used non-temperature specific drugs as pills for most of its purposes; however, the usage of liquid and other sterile products like injectable and infusion solutions has never been ceased either.

This is what a pharmaceutical company with the help of quality department has to do in order to maintain the product's quality throughout its transportation:

- They have to provide an ambient temperature, refrigeration control, and cold chain support
- The full chain of custody, that is from collection to handover signature, identity verification, and documented proof of the delivery, has to be demonstrated
- Having professional drivers who can offer safe and secure delivery without compromising on product safety, quality, and efficacy.

COLD CHAIN MANAGEMENT

It's not that all pharmaceutical products require a temperature controlled transportation. In fact, it is approximately over 30% of pharmaceutical products which require handling and transport at a controlled temperature so that they do not suffer any alteration. Depending on the drug or product, the temperature will be one or the other, and may range from negative to positive. For example, liquid nitrogen should be kept at −190 °C and frozen products at −20 °C. Drugs that require refrigeration will need temperatures between 2 °C and 8 °C, although there are also products that require room temperature, which is about 20 °C.

The cold chain management starts from the company's own warehouse where the released batches of products are stored until they are ready to go to the market or to the distributor's warehouse. There are various requirements that the QA needs to handle, given as follows:

1. Generate a Master plan of cold chain supply management.
2. Qualification of the company's stores for maintaining the environmental conditions like relative humidity and temperature.
3. Temperature mapping to determine the coldest and hottest points of the warehouse and installation of data loggers based on the mapping.
4. Qualification of the loading and offloading areas.
5. Usage and maintenance of calibrated data loggers.
6. Standard operating procedures for the functions stated above.
7. Qualification and temperature mapping of the transportation vehicles to determine the coldest and hottest points.
8. Inspection, maintenance, and monitoring of transportation vehicles. The medicines or pharma products are usually transported in a refrigerated or insulated truck. These trucks undergo routine inspections to make sure that they are under satisfactory conditions to keep inventory at the correct temperature.

QUALITY ASSURANCE OWNERSHIP

The Quality Assurance department has the ownership of maintaining the cold chain supply management standards.

It should not be forgotten that pharmaceutical transport is the most important part in a logistics chain because the products being transported are for human consumption. Therefore, it is necessary to have an exhaustive control of the whole distribution chain to ensure that the medicines arrive to the consumer under perfect conditions.

The manufacturer of pharmaceutical products is going to require the logistics operator that when handling and transporting this type of goods all the requirements are met. Safety, traceability, and temperature maintenance in pharmaceutical transport are critical elements that a logistics operator must guarantee.

Quality Assurance has the responsibility to have the data loggers installed in the transportation vehicles as per the qualification and temperature mapping results, monitor them, upload the data after the transportation process is accomplished and analyze them to draw conclusion.

COLD CHAIN REGULATIONS

Good Distribution Practice is an important regulation that is required in the for manufacturers and transport companies or suppliers of logistics services. This regulation guarantees safety and quality in the distribution of medicines, which are highly sensitive. Therefore, it regulates aspects such as equipment and installations, documentary follow-up, transport, vehicle cleaning, temperature monitoring, and inverse logistics.

QUESTIONS AND ANSWERS

1. **What is the significance of cold chain supply in pharmaceutical industry?**
   Answer: To maintain the pharmaceutical products' integrity and safety, the products must stay within the recommended ranges of temperature and humidity. The smallest temperature or humidity variation can alter the product's quality which can put the customer's health at risk and cost the pharmaceutical company a significant amount of money. In order to keep pharmaceutical products safe during transport, companies have to optimize their supply chain by ensuring that their goods are constantly stored under suitable conditions. To meet these requirements, companies that transport pharmaceuticals use a process known as cold chain shipping, or cold chain pharma logistics.

2. **Which department is the owner of the cold chain management standards?**
   Answer: The Quality Assurance department has the ownership of maintaining the cold chain supply management standards. Quality Assurance has the responsibility to have the data loggers installed in the transportation vehicles as per the qualification and temperature mapping results, monitor them, upload the data from the data loggers after the transportation process is accomplished, and analyze them to draw conclusion.

3. What are basic rules to maintain the products quality and integrity during logistics?

Answer: Pharmaceutical companies have to adapt the following basic rules in order to maintain the product's quality throughout the transportation;

- They have to provide an ambient temperature, refrigeration control, and cold chain support
- The full chain of custody, that is from collection to handover signature, identity verification and documented proof of the delivery, has to be demonstrated
- Having professional drivers who can offer safe and secure delivery without compromising on products safety, quality, and efficacy.

4. What are the functions that a pharmaceutical company should perform in order to maintain product's quality and during the logistics?

Answer: There are various requirements that the pharmaceutical company management must ensure under the guidance of QA. These requirements are given as follows:

1. Generate a master plan of cold chain supply management.
2. Qualification of the company's stores for maintaining the environmental conditions like relative humidity and temperature.
3. Temperature mapping to determine the coldest and hottest points of the warehouse and installation of data loggers based on the mapping.
4. Qualification of the loading and offloading areas.
5. Usage and maintenance of calibrated data loggers.
6. Standard operating procedures for the functions stated above.
7. Qualification and temperature mapping of the transportation vehicles to determine the coldest and hottest points.
8. Inspection, maintenance, and monitoring of transportation vehicles. The medicines or pharma products are usually transported in a refrigerated or insulated truck. These trucks undergo routine inspections to make sure that they are under satisfactory conditions to keep inventory at the correct temperature.

19 GMP for Biologics and Similar Medicines

WHAT ARE BIOLOGICAL AND BIOSIMILAR MEDICINES?

Biological drugs or biopharmaceuticals are made from proteins and other substances which exist in nature. To manufacture a biological product, a biological process is utilized instead of a chemical process. It can be easily understood that the biological drugs are a lot larger in size and more complicated than normal chemical drugs. Some common examples of such products are steroids, 5-ASAs, and azathioprine.

A biosimilar is also a biologic drug that is highly similar to a biologic drug that was already authorized for sale. The difference between biologics and biosimilar is same as that between a generic drug and an original brand product. There are no expected clinically meaningful differences in efficacy and safety between a biosimilar and the biologic drug.

However, in order for a biosimilar to gain approval to be used, it has to meet strict standards to show that it is as safe and effective as the original biologic drug, and to have the same quality, although it may have small structural differences.

PREPARATION OF BIOLOGICAL AND BIOSIMILAR DRUGS

Biological drugs are made using living organisms, such as modified plant and animal cells. The cells receive a gene (a piece of DNA) which instructs them to make a specific protein. The cell with the gene in it then multiplies, and produces the biological protein which can be "harvested". The cells that are the best at producing the protein are identified and grown on an industrial scale – these special cells are called a "cell line" and are unique to each manufacturer. The cell line is then grown in huge tanks, and the final product harvested. This is an extremely complex process, and very small changes to it can lead to very large changes in the final product. The biosimilar manufacturer must develop a new cell line and manufacturing parameters that result in a highly similar product.

GMP in Biologics/Biosimilar

GMPs prevent or minimize errors or defects during the manufacturing process, minimize contamination, and maintain the integrity and stability of biologic products just like all other pharmaceutical dosage forms. Therefore, the quality and safety of biologics are also very important, because the human health can be impacted in a number of ways by these products. Biologics are frequently sourced from other living things, and act with human metabolism to enhance an existing biological process. For example, vaccines enhance immune response, and hormones are administered

DOI: 10.1201/9781003366393-24

to patients with endocrine insufficiency. Therefore, it is essential that manufacturers adhere to strict good manufacturing practices in order to ensure product quality and safety, since failure to create a safe product has the potential to cause great harm.

Difference in GMP for Biologics and Other Dosage Forms

How do the good manufacturing practices for biologics or biosimilar products differ from solid dosage or any other pharmaceutical dosage form?

The answer to this question is not difficult. Although the biologics are different in nature and in manufacturing processes from other medical products and may have slightly different GMP requirements yet in a nutshell, the GMP does not significantly differ from that of other dosage forms. The reason is that the products, no matter manufactured in an entirely different way involving different types of processes instead of regular pharma products but still used to provide therapy to human beings. GMP is a system that assures that pharmaceutical and biological products are manufactured, sampled, and tested consistently in accordance with specific quality standards.

In case of biologics, the ancillary materials manufacturers can provide GMP quality by following all applicable GMP standards exactly the same way as advised for pharmaceutical products. To understand this further, in biologics manufacturing, ancillary materials are transfection reagents or a subset of critical raw materials which are used for gene therapy product manufacturing. Although they are raw ingredients that are not intended to be included in the finished product yet are vital in its manufacture. The requirements for the collection, testing, and quality control of ancillary materials are defined by a quality assurance system that is based on the existence of a national structure, independent of manufacturers and adheres to GMP principles and guidelines. Because the synthesis of biologically active substances and biological medicinal products for human use requires some specific considerations arising from the nature of the products and manufacturing processes, it is important to take some special precautions. Some AMs may be authorized medicinal products in their own right. Heparin and insulin are GMP-grade ingredients that are often utilized in cell product manufacture.

Cell- and tissue-based therapies are often manufactured in specialized facilities that contain equipment and methods for culturing, testing, and packaging live cells. Biologics are manufactured in a living system such as a microbe, plant, or animal cell. For cell product manufacturers, the nature of an Ancillary Material's production process is critical. Ancillary materials are also manufactured in accordance with all relevant GMP guidelines, documented verification of purity, potency, consistency, and stability. Therefore, GMP is one of the most important quality assurance systems for ensuring the safety and high quality of ancillary materials. Biologic products are typically more complex than traditional pharmaceuticals, due to the fact that they originate from living organisms which introduce variability due to genetics, environment, diseases, and impurities inherent to naturally-occurring biochemistry in the living thing. To understand the complexity of biologics, as an example, consider the case of a recombinant insulin product which may undergo a series of manufacturing steps, including fermentation to grow the cells generating the insulin, followed by another series of purification steps to obtain the desired product. A product which is derived from human blood may carry a risk of containing blood borne pathogens.

Best Practices for Manufacturing Biologics under GMP

Quality standards are evolving in response to current GMP rules. The original GMP regulations began with a focus on the product and quality control. The focus then shifted to procedure and quality assurance. With the new model of quality systems, pharmaceutical and biological firms are now facing a massive paradigm change. This GMP paradigm shift will change the fundamental assumption of the manufacturing process, radically shifting the regulatory and scientific emphasis by transferring the regulatory anchor from fixed processes to fixed (constant) outputs.

Following the revision of GMP regulations and the implementation of new technological advancements, the FDA anticipates that manufacturers will continuously maintain high-quality products and improve production efficiency. The FDA believes that this will help lower costs and minimize shortages of important medications caused by failures that may result in product recalls. It's indeed important to note that the FDA believes that the CGMP rules necessitate a higher level of calibration and maintenance than other non-pharmaceutical quality-system models.

Current GMP regulations will have a far greater impact on manufacturing and development.

Challenges in Biological Manufacturing under GMP

Since many AMs are complex biological materials, evaluating their qualities can be challenging. Biologics are difficult to work with since they are derived from living materials, which are naturally variable. As a result, the most difficult challenge from a GMP standpoint is ensuring consistent product quality, which is dependent on both incoming raw materials and manufacturing procedures. When additional complexity is included, such as a non-traditional dose form, consistency becomes increasingly challenging.

The complexity of biologics, as well as the application of emerging technology, offer challenges for cGMP compliance. In general, complying with the FDA's GMP requirements is not difficult, but there are several intricacies that might provide a barrier. For instance, it is simple to do gap analysis and confirm that processes are in accordance with the regulations, guidelines and technical publications comprising contemporary cGMPs. But still the complicated formulation work required in biological products is quite challenging.

Risks in Biological Product Manufacturing and Mitigation

It is evident that biologic products have numerous risks resulting from their inherent nature. Due to this complexity and risk many additional GMP requirements are needed to be met to achieve GMP compliance. For example, manufacturing facilities used to produce biologics must be designed and constructed specifically for this purpose. In addition, all personnel working in these facilities must undergo rigorous training. For example, personnel should not go to work in other areas outside the biologics production facility in the same working day. In the following pages, let us review the particular areas where specific precautions are to be taken to follow the GMP and maintain compliance.

FACILITIES AND EQUIPMENT

The following measures will be critical for GMP;

1. The design, construction, and maintenance of manufacturing facilities and equipment should be in such a way that may minimize the risk of contamination.
2. All production areas must be dedicated to prevent cross-contamination with other processes.
3. Some equipment for biologics manufacturing may be highly-specialized, and will require special testing to ensure that its use is suitable for making biologic products.

PERSONNEL

1. All personnel working in manufacturing facilities must be properly trained and qualified.
2. They must also follow all procedures and protocols for clothing requirements to minimize the risk of contamination.
3. Health monitoring process should be in place and strictly adhered to, to prevent disease crossover into the product.

MATERIALS

1. The materials used in the manufacturing process must be of high quality and free of contaminants.
2. All sterilization cycles are validated.
3. Specialized gowning process is followed.
4. Materials are segregated to ensure that they retain their purity through manufacturing, holding, and distribution.

QUALITY CONTROL

Needless to say, Quality Control testing for biologic products is extensive, from raw material receipt through the end product. Laboratory testing is just part of a comprehensive quality control program.

1. There must be a strict quality control program in place to ensure that all products meet the required standards, including microbial control, standards for animal care, holding retain samples 6 months past expiration, and temperature control during shipment.
2. The FDA requires reporting of deviations. Therefore all biological product deviations are required to be investigated.

Good Laboratory Practices in Biologics Manufacturing

The safety and quality of biologic products are ensured by the GMP guidelines no matter how overwhelming they may seem, they are essential. By following these guidelines, manufacturers can be confident that their products will meet the highest standards of quality and safety.

- Quality assurance work in biologics/biosimilar is very much like in other pharmaceutical manufacturing, that involves assuring and verifying that controls in the design of the biologic facility continue to perform as designed.
- Apart from that, after release stability testing is also required.
- Quality Control also includes validation of laboratory testing as well as the area where biologic material is harvested from animals, with specified requirement for assuring cleanliness, including assurance of animal health and preventing incursion of spore-forming microorganisms.
- Laboratory test methods also require validation in much the same way that validation is required for non-biologic pharmaceuticals products, but the specific laboratory tests required will differ.
- The types of test methods used for testing biologic products including
    - In-vitro assays
    - Immunoassays
    - Chromatography
- The choice of testing method will depend on the type of product being produced, the scale of production, and regulatory requirements.
- Regardless of the testing method used, it is essential to ensure that all testing is carried out in a way that minimizes the risk of contamination, the equipment used in testing is qualified, and the test methods are validated to provide a true, reliable result.
- Equipment sterilization requirements are specified in the standard, and sterility via equipment handling procedures to assure continued sterility.
- Tests may include screening for specified animal diseases, and reporting certain diseases to the FDA if an outbreak is detected.

  It is important for manufacturers to stay up-to-date with the current GMP requirements. Failure to do so can result in subpar products and potential safety risks for consumers, or products failing to meet the expected medical state-of-the-art.

QUESTIONS AND ANSWERS

1. **What is a biosimilar drug?**
   Answer: A biosimilar drug is also a biologic drug that is highly similar to a biologic drug that was already authorized for sale. The difference between biologics and biosimilar is same as that between a generic drug and an original brand product. There are no expected clinically meaningful differences in efficacy and safety between a biosimilar and the biologic drug.

However, in order for a biosimilar to gain approval to be used, it has to meet strict standards to show it is as safe and effective as the original biologic drug, and to have the same quality, although it may have small structural differences.

2. **In what respect GMP is important in Biologics/Biosimilar drugs manufacturing?**
   Answer: Good Manufacturing Practices prevent or minimize errors or defects during the manufacturing process, minimize contamination, and maintain the integrity and stability of the products just like all other pharmaceutical dosage forms. Therefore, the quality and safety of biologics are also very important, because the human health can be impacted in a number of ways by these products. Biologics are frequently sourced from other living things, and act with human metabolism to enhance an existing biological process. For example, vaccines enhance immune response, and hormones are administered to patients with endocrine insufficiency. Therefore, it is essential that manufacturers adhere to strict good manufacturing practices in order to ensure product quality and safety, since failure to create a safe product has the potential to cause great harm.

3. **Describe Good Laboratory Practices in Biologics Manufacturing.**
   Answer: The GMP guideline work in biologics/biosimilar is very much similar to other pharmaceutical manufacturing that involves assuring and verifying that controls in the design of the biologic facility continue to perform as designed. Apart from that, after release stability testing is also required.

   Quality Control also includes validation of laboratory testing as well as the area where biologic material is harvested from animals, with specified requirement for assuring cleanliness, including assurance of animal health and preventing incursion of spore-forming microorganisms.

20 QA/QC Release

Some of the responsibilities and functions discussed in this chapter are already mentioned, though briefly, in Chapter 17. In this chapter those very functions will come under detailed discussion in the context of batch release.

The Quality Control (QC) department is critical in ensuring pharmaceutical products' safety, efficacy, and quality in the pharmaceutical industry. QC releases raw materials, packaging components, in-process samples, and finished products.

The QC department ensures that these raw materials and packaging components meet the required quality standards before they are used for production.

These materials are received and stored in the warehouse in a controlled environment.

Raw materials are segregated based on their characteristics, such as active pharmaceutical ingredients (APIs), excipients, solvents, or packaging materials.

If an investigation is required due to any reason, the material is kept in the quarantine area with the sticker "quarantine" signed by the QA official. Similarly, if the material is rejected, the material is kept in a rejected area with the sticker "rejected" signed by the quality official and later disposed as per policy.

Each raw material as well as finished product batch should be labelled with relevant information such as name, batch number, manufacturing and expiration date, and supplier details.

Records of incoming and outgoing materials, including certificates of analysis, should be maintained. Good documentation practice is essential to maintain traceability and regulatory compliance.

The containers should be opened, sampled, and closed to prevent contamination of their contents and other components.

Additional safety measures should be taken if the raw materials include hazardous substances.

Sterile equipment and aseptic sampling techniques shall be used when necessary.

These include proper labelling, storage in designated areas with appropriate safety equipment, and adherence to regulations regarding the handling and disposal of hazardous materials.

Materials received first are used first. Those materials are used first, which have less expiry period than others.

RECEIVING AND RELEASE OF RAW AND PACKAGING MATERIALS

Releasing raw materials and packaging components is critical to ensure the final product's quality, safety, and efficacy in the pharmaceutical industry. The quality control (QC) department plays a key role in this process.

DOI: 10.1201/9781003366393-25

Let us take a look of the entire process as to how the Raw, packaging materials and active pharmaceutical ingredients and excipients are received, stored, tested, and then released in the QC Laboratory.

Packaging pharmaceutical products are bottles, blister packs, labels, inserts, and outer cartons.

A representative sample is collected from the warehouse. Sampling methods follow established guidelines and procedures to ensure that the sample accurately represents the bulk material.

Personnel working in manufacturing, processing, packing, or holding a drug product should wear clean clothing appropriate for their duties. Personal protective equipment, such as head, face, hand, and arm coverings, should be worn as necessary to protect drug products from contamination. Personnel shall practice good sanitation and health habits.

TESTING OF RAW MATERIAL

Quality Assurance personnel will collect and submit the sample to the QC department. The submission has to be documented with the date and time the sample was submitted. QC officials arrange required testing for materials.

The QA team reviews the documentation the supplier provided, including a certificate of analysis (CoA) that describes the test results and quality attributes of the raw material.

The QA team verifies the accuracy and completeness of the documentation.

The sample undergoes tests and analyses to evaluate its quality and compliance with predefined specifications. These tests may include identification, physical, chemical, microbiological, and other relevant tests for the specific material and its intended use.

THE FLOW OF QC WORK

After the QA personnel submits the samples to the quality control laboratory, the flow of work occurs as discussed below;

Samples Entry

The sample must be labelled with the name of the sample, lot number, and test required. The sample is entered in a logbook or register. The sample identity, time, and date of the sample submitted are entered in the logbook.

Planning

The laboratory supervisor or manager with other departments, such as Quality assurance or Production, determines the upcoming workload and prioritize tasks accordingly. This planning helps establish the overall schedule for the lab and ensures that critical tasks are addressed promptly.

In some companies, specific departments or groups deal with the samples of raw materials, packaging components and in-process testing separately.

In-process testing is always prioritized since the production waits to proceed to the next step of production.

Work Orders/Batch Records Creation

Once the work plan is established, specific work orders or batch records are created for each task or project. These documents outline the required tests, methods, specifications, and other relevant information for the analysis or quality control activities.

Work Assignment to Analysts

The lab supervisor, manager, or designee assigns the work orders to individual lab analysts or technicians based on their skills, expertise, and workload capacity. Consideration is given to balancing the workload among team members and ensuring all tasks are completed within the required timeframes.

The lab supervisor, manager, or designee create a sample request, which contains the list of test to be performed and the method identification for each test. It also contains the name of the analyst performing each test. Each analyst enters his name, the date testing started, and date testing finished.

Before assigning tasks, it is crucial to ensure that lab personnel are properly trained and qualified for the specific tests or analyses they perform. This includes comprehensive training on test methods, equipment operation, safety protocols, and Good Laboratory Practices (GLP) or Good Manufacturing Practices (GMP) requirements.

Lab analysts receive clear instructions and guidelines regarding the assigned tasks. These instructions may include details about sample preparation, testing procedures, instrument usage, data recording, and any specific quality control requirements. Effective communication ensures that analysts understand their responsibilities and perform their work accurately.

Analysts maintain good documentation practice throughout the process, including observations, test results, deviations, and other relevant information. Accurate record-keeping is vital for traceability, quality assurance, and regulatory compliance.

In a quality control lab, internal quality control checks are performed to ensure the accuracy and reliability of test results. This involves verifying the performance of equipment, calibrating instruments, running control samples, and participating in proficiency testing programs to validate the lab's competency.

Results Review

Once the assigned work is completed, lab analysts submit their findings and data to the appropriate authority, such as a senior scientist, quality manager, or quality assurance team. These individuals review the results, verify compliance with specifications and protocols, and approve the data before it can be used for decision-making or further processes.

Laboratory Information Management System (LIMS)

Some companies use the Laboratory Information Management System (LIMS), which is a software-based solution that helps manage and track laboratory operations and data. LIMS is designed to streamline and automate laboratory processes, enhance efficiency, improve data integrity, and ensure regulatory compliance.

LIMS facilitates sample tracking throughout the laboratory workflow. It assigns unique identifiers to each sample, records sample information and track its movement from one laboratory area to another. This ensures samples are properly identified and prevents mix-ups.

LIMS helps define and manage laboratory workflows, ensuring that tasks and analyses are performed according to standard operating procedures (SOPs). It provides a centralized platform for assigning and tracking work, ensuring proper sequencing of tasks, and monitoring progress at each stage. This improves efficiency and reduces errors by eliminating manual paperwork and reducing human intervention.

LIMS acts as a centralized database for storing and managing laboratory data. It captures and stores raw data, test results, and associated data generated during laboratory analysis. This includes information on instruments used, reagents, standards, calibration, and quality control measures. LIMS provides secure data storage, version control, and audit trails to ensure data integrity and traceability.

LIMS can integrate with various laboratory instruments and equipment, allowing data transfer and automation of analysis workflows. It eliminates manual data entry, reduces transcription errors, and improves efficiency. Instruments can directly communicate with LIMS to transmit data, initiate tests, and receive instructions, streamlining the analytical process.

LIMS offers data analysis and reporting capabilities to support decision-making and generate actionable insights. It can perform calculations and statistical analysis and generate reports with customizable formats. LIMS can also provide trend analysis, generate graphs, and facilitate data visualization for easier interpretation and comparison of results.

LIMS enables collaboration and information sharing across different organizational departments or laboratories. It supports data exchange, remote access, and electronic approvals, promoting efficient communication and collaboration among scientists, managers, and quality assurance personnel.

Depending on the results obtained, further actions may be required, such as investigation of out-of-specification results, troubleshooting equipment issues, or reporting deviations to the quality assurance department. Lab analysts may be involved in these follow-up activities to resolve any identified issues.

It's important to note that the specific processes and workflow may vary between pharmaceutical companies, depending on their size, organizational structure, and the nature of the work. However, the underlying principles of planning, assignment, documentation, and quality control remain consistent to ensure the integrity and reliability of laboratory work in the pharmaceutical industry.

TESTING OF PACKAGING COMPONENTS

Packaging components are also submitted for testing. The QC department verifies the quantity and inspects the components for visible defects or damage. This includes checking the integrity of the packaging materials, ensuring correct dimensions, and examining for contamination.

Various tests are performed on these samples to evaluate their quality. These tests include dimensional checks, visual inspections, functional assessments, and verification of printed information such as barcodes, batch numbers, and expiry dates. The components may also undergo specific tests, such as container leak tests or ink adhesion tests for labels.

Inserts, labels, blisters, and primary cartons contain specific text. This text is verified by comparing it with standards. The specific color is also matched with the standard. Those standards are kept on record.

APPROVAL OF PACKAGING COMPONENTS AND RAW MATERIALS AFTER TESTING

The test results obtained in the laboratory are compared against the acceptance criteria and specifications for the raw material. The QC team evaluates whether the material meets the required standards.

After comparing the test results with specifications, documentation review, and compliance assessment, the QA department decides on the release of the raw material. If the material meets all specifications and quality standards, it is considered suitable for use in the production of pharmaceutical products.

The necessary documentation is released to record the decision when the raw material is approved. This may include a release certificate or a batch record indicating that the material has passed all necessary tests and is approved for use in production.

Once the material components are approved, they are appropriately labelled, identified, and stored in a controlled environment to prevent mix-ups or contamination until they are needed for production.

In case of any non-conformances or deviations, an investigation is carried out, and appropriate actions are taken, such as rejecting the batch or initiating corrective actions.

RELEASE OF IN-PROCESS AND FINISHED PRODUCTS

In-process quality control tests are simply routine tests performed during different stages of production. These tests are carried out during the manufacturing process.

These tests ensure that product quality is met before the final product is manufactured.

This may include control of equipment and environment also.

The objective of in-process testing is as follows:

- To optimize the technical procedure used in the manufacturing process.
- To monitor, control, and improve the applied operations effectively at every stage of the finished pharmaceutical products.
- Inspection of the raw material, equipment, environment, process, and testing with respect to specification, packing, etc.
- Quality and process control.

In-process quality control is performed regularly during a process step or at the end.

These are checks that are carried out before the manufacturing process is completed. The function of in-process controls is monitoring the manufacturing process to comply with the specifications. This may include control of equipment and the environment too.

In-process materials should be tested for identity, strength, quality, and purity as appropriate and approved or rejected by the quality control unit during production. Rejected in-process materials should be identified and controlled under a quarantine system designed to prevent their use in manufacturing.

Common tests performed during in-process testing are as follows.

Moisture/Loss on Drying Test at the Granulation Stage

After granulation, tablets are compressed. For this purpose, a certain moisture is required. Before tablets are compressed production team wants to ensure that moisture is optimum and granules are ready to be compressed.

pH

For the liquids, pH is important. This test determines whether the solution is acidic or basic and whether its acidity/basicity complies with the specification.

pH affects the stability of the product, so before filling and after filling of suspension and syrup pH has to be checked out for product consistency.

Weight Variation

This test is performed during the compression of tablets to ensure that the weight of individual tablets is as per specification.

Hardness

Monitoring tablet hardness is especially important for drug products with potential bioavailability problems sensitive to altered dissolution release profiles as a function of the compressive force employed.

Disintegration

A tablet form of the drug should be readily available to the body and must be in solution. For most tablets, the first important step toward a solution is the breakdown of the tablet into smaller particles or granules, a process known as disintegration.

Content Uniformity

A physically sound tablet may not produce the desired effects.

To evaluate the efficacy of a tablet or capsule, the amount of drug per tablet needs to be monitored from tablet to tablet, and a measure of the ability of the tablet to release the drug needs to be ascertained.

Assay of Active Ingredients

The amount of an active pharmaceutical ingredient in any dosage form is called an assay. This is one of the most important tests performed on finished products and different stages of the in-process.

Active ingredient means pure drug present in the product. An assay of an active ingredient must be done because it is the only one responsible for pharmacological action.

TESTING

Samples are tested to verify conformance with specifications such as identity, components conformity to written specifications, container/closure conformity with written specifications, and examination for contamination. The tests to be performed and the methods to be used must be defined and specifications established.

QA notifies production if the results comply, and production may proceed to the next step.

The same tests are performed on the finished product, too, depending upon the monograph of the product.

Tests on Finished Products

Other than the above tests, the following tests are also performed on finished products.

Dissolution Test

Dissolution testing measures the rate at which a drug substance dissolves in a specific solvent or medium, which is directly related to the drug's bioavailability.

Related Substances

Related substance tests measure the quantity of degradation of active compounds in the product.

Random samples are collected from the batch or lot of finished products according to predetermined sampling plans or guidelines by quality assurance.

The sample is submitted for testing to QC. This submission is documented with sample identification and date and time of submission.

Out of Specification Results and Lab Investigation Report (LIR)

After completing a test, if the analyst finds any aberrant result, the analyst must inform management immediately to initiate a lab investigation report (LIR).

There are two types of aberrant results:

- Out-of-specification (OOS) results.
- Out-of-trend (OOT) results.

Investigating out-of-specification (OOS) results in a pharmaceutical laboratory is critical to ensure the quality and integrity of pharmaceutical products.

Here is a general outline of how to conduct investigations:

When an aberrant result is generated, and a verification/review of the raw data identifies a wrong manual calculation or incorrectly manually entered parameters in a software calculation as the cause, LIR is not required to be opened.

If, after review, the OOS/OOT is confirmed, LIR must be initiated. This can happen during routine testing, quality control, or release testing of pharmaceutical products.

The OOS result is documented, including relevant details such as sample information, test method, equipment used, and any observations or deviations from standard procedures. The laboratory notifies the appropriate personnel, such as the laboratory supervisor, quality assurance (QA) department, and/or the manufacturing site.

The laboratory investigation process should be designed to ensure the proper sequence of actions and controls in the investigation of any unusual laboratory results to assess product quality impact and determine the root cause of any errors that occurred during the testing or manufacturing of the product.

Initial Lab Assessment

The laboratory conducts an initial assessment to determine the criticality of the OOS result. This involves reviewing relevant data and historical records and assessing the impact on patient safety, product quality, and regulatory compliance.

The laboratory performs an initial investigation to determine if the OOS result is due to an outlier or a laboratory error.

First, a bench audit is done to verify that the correct glassware, sample, and instrument parameter are used.

This may also involve retesting the original sample, verifying the test procedure, and evaluating the equipment calibration and maintenance records.

A detailed investigation plan is developed if the OOS result is confirmed and not due to laboratory error. The plan outlines the scope, objectives, and investigation steps. It may include thoroughly reviewing batch records, sample history, manufacturing processes, and potential root causes.

The purpose of the laboratory investigation phases is to systematically determine the root cause of the aberrant result so that the validity of the analytical result can be evaluated, and product quality impact can be assessed.

All analytical results shall be determined and reported as specified in the analytical method. If more than one sample preparation injection is required for the analysis, each result should be evaluated for compliance with the specification before averaging. Any individual OOS result should be subject to the investigation process.

If the aberrant result occurs on accelerated or intermediate stability studies, the extended investigation may not be required; however, the rationale must be documented in the laboratory investigation report along with a complete assessment of the potential impact on the distributed product.

All standard and sample preparation, glassware, reagents, and instrumentation must be retained and reserved to aid the investigation.

All data acquired during an investigation, even if invalidated, shall be retained with the data of record and an electronically approved copy of the LIR.

During the investigation, all actions must be justified based on sound scientific judgment, documented, and approved before performing each action.

The scope of the error should be evaluated and justified in determining which samples should be tested as part of the evaluation. The original analyst may perform the additional testing, or a second analyst may perform it. The identity of the analyst responsible for the additional testing will be documented.

The investigation may consist of the following steps.

Re-injection/Re-scan

If an instrument malfunction or external contamination of the sample vial is suspected and the samples are within their stability, the analyst will re-injection/re-scan the original laboratory working sample solution from the original vial and a new vial. Suppose the assignable cause is determined to be an instrument malfunction or external contamination. In that case, the re-injections/re-scans of the sample solution will stand as data of record provided that the samples are within the established time frame for solution stability and all system suitability criteria are met. If an instrument malfunction affecting more than one reading/injection is suspected, management may initiate reinjection/reread of all solutions.

Re-dilution

If pipetting of the original laboratory stock sample/standard solution is suspected, the analyst will re-dilute the original laboratory stock sample solution following the testing method, with no additional extraction of the sample, and the vial with the re-diluted solution will be read together with a re-read of the initial vial. If re-dilution discovers the original aberrant result, the results of the re-dilutions can be considered valid and used as the reportable results, provided the re-dilutions were performed within the solution stability and all system suitability criteria were met. If re-dilution discovers the original aberrant result but is performed outside the solution stability, a repeat test may be performed on the sample/lot to confirm that the original results were due to a dilution error.

Re-extraction

If an error in the extraction of the drug substance or homogenization of the sample solution is suspected, the analyst will re-analyze the original laboratory stock sample/standard solution. Suppose the re-analysis discovers the original aberrant result. In that case, the original results may be invalidated due to insufficient extraction, and a repeat test may be performed on the sample/lot. The results of this exercise must support the assignable cause theory and be substantiated with good scientific rationale and observations.

Standard Study

If an error in standard preparation is suspected, a standard study will be initiated. A standard check may be performed using multiple preparations/dilutions/ injections of freshly prepared standards and the original standard solution.

Testing Resolution

Reassessment testing is evaluated and dispositioned at the end of each step. If a clear assignable cause has been determined and valid results were obtained through re-injection, re-dilution, or re-extraction, the original result may be invalidated with a clear written explanation in the LIR.

The valid results obtained during the above actions can be considered reportable, provided that the results of this exercise support the assignable cause theory and are substantiated with good scientific rationale and observations. No further investigating step is required.

Re-testing

If the above steps do not discover any cause, retesting should be performed using the original laboratory sample. The number of sample solutions prepared for any retest should be specified in the LIR record and be justified and sufficient to provide sufficient relevant data. The retest can be performed by a second analyst who is at least as experienced and qualified in performing the method as the original analyst or by both the original analyst and a second analyst.

Re-sampling

Re-sampling is performed only when the original sample is exhausted or degraded. Re-sampling can also be performed when the initial investigation identifies a potential sampling error or the sample obtained is not homogenous or representative of the lot/batch. A solid scientific rationale is required.

Extended Investigation

If the above investigation does not identify a determinate laboratory error, the initial results are considered valid, and the next step of the investigation is required.

This investigation should be initiated promptly and carried out concurrently with the extended laboratory investigation to limit the continuance or carryover of a potential manufacturing deviation to produce lots subsequently.

A review of production records, batch handling, in-process testing, operating ranges/times, etc., will be performed to determine if any process deviations may have resulted in the aberrant result.

ROOT CAUSE ANALYSIS AND CORRECTIVE AND PREVENTIVE ACTIONS

The laboratory conducts a comprehensive root cause analysis (RCA) to identify the factors that led to the OOS result. This involves analyzing data, reviewing laboratory practices, investigating potential equipment failures, assessing environmental factors, and considering personnel-related issues.

Based on the identified root causes, the laboratory develops a set of corrective and preventive actions (CAPA) to address the issues and prevent their recurrence. CAPAs may involve equipment calibration, procedural revisions, training programs, environmental monitoring, or corrective actions.

The laboratory implements the identified CAPAs and verifies their effectiveness. This may involve retesting samples, performing method validation or verification, and conducting equipment qualification or any other appropriate measures to ensure the CAPAs have resolved the issue.

All activities, findings, and actions are thoroughly documented throughout the investigation. A final report summarizing the investigation steps, results, and conclusions is prepared. The report also includes any preventive measures implemented and their expected impact on future operations.

The investigation report is reviewed by relevant personnel, such as the laboratory supervisor, QA department, and/or the manufacturing site.

CONCLUSION OF AN LIR

The effecting batch is kept on hold until the final approval of LIR from the QA. If the LIR reveals that the OOS result was due to an analytical error and the product is not impacted, the batch is released. If LIR reveals no error has occurred and the production batch is not meeting specification the batch will be rejected.

QC analyses to evaluate their quality, including physical, chemical, and microbiological tests. These tests are performed using validated methods and equipment.

Throughout the finished product release and in-process release processes, strict adherence to good manufacturing practices (GMP) and regulatory guidelines is maintained to ensure the quality and safety of pharmaceutical products. Quality control professionals, often working in dedicated laboratories, play a vital role in executing these processes and ensuring that only products meeting the required quality standards are released for distribution and sale.

RELEASE OF A PRODUCTION BATCH

Upon receipt of the batch records (batch manufacturing and batch packing records), Quality Assurance will review the documents for any discrepancy or deficiency which impacts the product quality and customer safety.

The pharmaceutical product follows validated manufacturing procedures and guidelines. This includes the active ingredients formulation, dosage form preparation, and packaging.

Various quality control in-process tests are performed during manufacturing to verify the product's identity, potency, purity, and overall quality. These tests are conducted in accordance with regulatory standards and may include physical, chemical, and microbiological analyses.

Detailed records are maintained at each manufacturing stage, documenting the processes, materials used, and quality control test results. This documentation is crucial for batch traceability and regulatory compliance.

Once the manufacturing and quality control testing is completed, the batch undergoes a thorough review by qualified quality personnel, such as a quality assurance team or a designated authorized person. They examine the batch records, test results, and any deviations or investigations during production.

QA personnel shall review and verify the following points before preparing the release order:

- Ensure that production has reviewed and audited the batch manufacturing record in all aspects.
- The reconciliation of the materials and product yield is carried out at the respective stages of the operation.
- Ensure that Production Manager has reviewed the completed batch manufacturing record before it is submitted to Quality Assurance for final audit and release of the Finished Product for distribution along with the product Transfer Note.
- Review the batch manufacturing record.
- Verify that the product has been approved by Quality control concerning its testing as per the product release specification.
- Ensure the Certificate of Analysis (COA) and analysis record have been signed.
- Attach the Certificate of Analysis (COA) and the batch manufacturing record; make the signatures and date at specified places.
- Verify the quantity being released from the quantity transferred from the warehouse to the production.

Before releasing the batch, it is important to ensure compliance with applicable regulations, such as Good Manufacturing Practices (GMP) and other regularity authorities.

The responsible personnel authorize the batch release for distribution and sale upon completing the batch review and regulatory compliance checks. This signifies that the batch has met all the necessary quality standards and is deemed safe and effective for use by patients.

The release of a production batch is accompanied by the generation of appropriate documentation, including a batch release certificate. This document provides information about the batch, its manufacturing details, and quality control test results as evidence of compliance and traceability.

Any discrepancies that directly impact the quality of the product or the consumer's safety and any discrepancy that leads to batch rejection/ failure/ serious non-compliance to approved procedures shall be classified as a critical discrepancy.

These discrepancies will be brought to the notice of Head-Quality.

These discrepancies may arise from failure to carry out GMP/GLP, etc.

All critical discrepancies observed should be investigated or corrected in consultation and agreement with the Head of Quality with the Head of Production.

The release of a batch should be based on the investigation findings/ closure of CAPA/corrective action, and the Head of Quality shall make the final decision.

The responsible personnel authorize the batch release for distribution and sale upon completing the batch review and regulatory compliance checks. This signifies that the batch has met all the necessary quality standards and is deemed safe and effective for use by patients.

Based on the review, a release decision is made by the quality control unit. If the product meets all the predefined specifications and quality standards, it is deemed suitable for release into the market.

QUESTIONS AND ANSWERS

1. **What is the role of a warehouse in the pharmaceutical industry?**
   Answer: Raw materials and packaging components are received and stored in the warehouse in a controlled environment.

2. **Which department is responsible for sampling raw materials, packaging components and finished products?**
   Answer: Quality assurance (QA).

3. **Which department is responsible for testing raw materials, packaging components and finished products?**
   Answer: Quality Control (QC).

4. **What material is classified as raw material?**
   Answer: Active pharmaceutical ingredients (APIs), excipients, and solvents.

5. **What is in-process testing?**
   Answer: In-process testing is done during different stages of production. Production must wait for an in-process result to proceed to the next step.

6. **If a raw material is approved, what is being done to the material in the warehouse?**
   Answer: It is transferred to the "Approved" section of the warehouse with an approved sticker signed by a quality official.

7. **What should be included in a label of raw material?**
   Answer: Each raw material should be labelled with relevant information such as name, batch number, expiration date, and supplier details.

8. **During the review, what will be the next step if any result is out of specification?**
   Answer: An investigation will be opened, and the batch will be rejected. The corrective and preventive action will be determined to prevent it again.

9. **During the review, what will be the next step if all results comply with the specification?**
   Answer: The batch will be released and sent to distribution.

10. **What are the packaging components?**
    Answer: Packaging pharmaceutical products are bottles, blister packs, labels, inserts, and outer cartons.

11. **What are the packaging components?**
    Answer: Packaging pharmaceutical products are bottles, blister packs, labels, inserts, and outer cartons.

12. **What tests are performed on packaging components?**
    Answer: These tests include dimensional checks, visual inspections, functional assessments, and verification of printed information such as barcodes, batch numbers, and expiry dates. The components may also undergo specific tests, such as container leak tests or ink adhesion tests for labels.

13. **What common tests are being performed during in-process testing?**
    Answer: Moisture/Loss on drying test at the granulation stage, pH, Weight variation, Hardness, Disintegration, Content uniformity and Assay of active ingredients.

14. **What are the objectives of in-process testing?**
    Answer: (1) To optimize the technical procedure used in the manufacturing process. (2) To monitor, control, and improve the applied operations effectively at every stage of the finished pharmaceutical products. (3) Inspection of raw material, equipment, environment, process, and testing with respect to specification, packing, etc. (4) Quality and process control.

15. **What should be an analyst's action if he finds an OOS result?**
    Answer: Inform management immediately.

16. **If an LIR is in the process, what would be the status of the effecting batch?**
    Answer: It will be on hold until the LIR is closed.

17. **During an investigation, when is the re-sampling required?**
    Answer: Re-sampling is performed only when the original sample is exhausted or degraded. Re-sampling can also be performed when the initial investigation identifies a potential sampling error or the sample obtained is not homogenous or representative of the lot/batch. A solid scientific rationale is required.

Index

aberrant result 243–245
accelerated studies 183, 185, 188
accelerated testing 182–184
acceptable quality limit 89
acceptance criteria 36, 43, 70, 82–84, 87, 96, 107, 108, 110, 115, 123, 135, 136, 157, 158, 186, 219
ACS 21, 28, 57
active pharmaceutical ingredient 32, 220, 237, 248
ALCOA 63–67
ambient condition 187
ambient temperature 227, 229
American Chemical Society 28
anhydrous basis 28
application screen flow 178
approved procedure 5, 247
AQL 82, 89, 107
aseptic manufacturing 199, 204
atomic absorption 29, 123
attributable 63, 64, 66, 67, 202
autoclaving 203
autoinjector 19

backpressure 127
benchmarking 216
bill of material, (BOM) 173, 193, 195
bioavailability 231, 232
biologics 25, 230–235
biosimilar 65, 230, 231, 234, 235
blend uniformity 77, 87, 88, 91, 94, 99
blending 77, 78, 85, 87, 91, 94, 99, 100, 102, 198, 201, 207
blistering 81, 85, 95, 106, 112
bracketing 183
buffer solution 21, 29, 30, 54, 55, 57

CAPA 8, 221, 246, 248
capacity factor 130
certificate of analysis 9, 20, 57, 220, 247
change control system 11, 221
change control tracking 12
chromatography 123, 126–128, 130, 234
chromophore 127
clean in place (CIP) 118
clean out of place, (COP) 118, 125
cleaning agent 121–123, 125
cleaning agents grouping 121, 122, 123, 125
cleaning methods grouping 121
cleaning validation 9, 117–125
cold chain supply 226–229
cold chain support 227, 229
colorimetric solution 29
column 17, 126–129, 132, 133
column heater 17, 128, 133
commercial manufacturing 3, 12, 189
commercial product 36, 71, 93, 115, 189, 191, 220
commercial production 71, 93, 115, 189, 191
composite sample 32, 33
compression 77, 80, 85, 89, 95, 99, 103, 106, 112, 183, 198, 207, 241
concurrent validation 70–72
constant weight 49
container closure system 184
contemporaneously 63
content uniformity 32, 34, 47, 88, 103, 114, 203, 242, 249
context diagram 178
control air 170
corrective and preventive action 221, 246
correlation coefficient 131
critical alarm 165, 166
critical quality attribute 35, 69, 70, 72, 115
cross-contamination 194, 215, 217

data flow diagram 178, 180
data integrity 6, 63–67, 239
data logger 227–229
data security 63, 67
degradate 182, 183, 185, 187
degradation product 35, 36, 131, 132, 183–187
design qualification (DQ) 134, 135, 139
design specification 69, 72, 156, 173
detection limit 126, 132
detector noise 127
diluent 129
diluting solvent 32, 37, 38, 129
direct surface monitoring 123, 125
disintegration 39, 51–54, 61, 77, 80, 99, 103, 242, 249
dispensing 74, 193–197, 214
dispersible tablet 54, 61
dissolution 19, 20, 23, 38–46, 48, 103, 104, 106, 108, 109, 113, 114, 221, 225, 241, 242
DNA 230
dosage form 32, 39, 40, 41, 47, 76, 89, 98, 106, 121, 124, 126, 198, 199, 202, 206, 207, 210, 226, 230, 231, 235, 242, 246
dried basis 28
dried granule 79, 102
drug substance 35, 36, 184, 187

electrode 20–23, 55–57, 61
electronic record 6, 66, 219
elemental impurities 35
emergency stop 165
encapsulator 198, 206
equipment grouping 121
equipment train 119
excipient 6, 32, 36, 76, 77, 88, 98, 99, 120, 131, 198, 220, 236, 237, 248
expiry date 21, 55, 85, 96, 112, 182, 188, 193, 211, 240, 249
extended investigation 244, 245
external contamination 244

FDA 11, 13, 14, 63, 68, 71, 72, 93, 139, 216, 219, 232–234
film-coated tablet 41, 44
flame photometry 29
flow rate 17, 18, 33, 37, 126–128, 132, 146–148
fluorescence detector 127
force degradation 130, 132, 133, 187
formal stability studies 183, 184, 188
friabilator 16, 23, 59, 60
friability 16, 60, 62, 77, 99, 113
functional specification 165, 173, 175–179, 181

GDP 1, 3, 5, 7, 9, 11, 13, 64, 222
glass electrode 20, 55
good documentation practice 1
good laboratory practice 1, 5, 27, 234, 235, 238
good manufacturing practice 1, 4, 12, 68, 69, 72, 193, 218, 219, 231, 235, 238, 246, 247
granulating 77, 78, 95, 99, 100, 101
granulation 78, 94, 101, 207, 241, 249

hard capsule 55
hardness 58, 59, 62, 77, 80, 99, 103, 241, 249
health regulatory bodies 117, 226
HEPA filter 195
historical record 243
homogeneity 87, 90, 96, 201
homogenization 245
HPLC 17–19, 23, 32, 33, 37, 44–46, 123, 126–130, 132, 219, 225
hydrogen ion 20, 29, 30, 50, 55
hydrophobicity 121

ICH guidelines 68
immunoassays 234
impurity standard 25
indicator 29, 30, 166
injection volume 33, 37, 126, 128
initial lab assessment 243
inorganic impurities 35
inorganic salt 35
in-process 49, 55, 81, 85, 90, 103, 201, 203, 220, 225, 241, 245
in-process control 3, 74, 97, 225, 241
installation qualification (IQ) 134, 136–141
instrument malfunction 244
interface diagram 178, 180
intermediate product 3, 224
intermediate storage condition 184, 185
international standard (IS) 25, 216
in-vitro assay 234
ion-pair reagent 126, 128
Ishikawa diagram 216, 217
ISO 168, 216

Karl Fischer (KF) 50
known impurity 37

lab investigation report (LIR) 243
laminar flow hood 195
leak test 81, 106. 240, 249
lifecycle 3, 12, 27, 64–66, 176, 215, 222, 225
LIMS 7, 239
linear regression 19, 131, 186
linearity 18, 19, 126, 130, 131, 133
liquid-liquid extraction 129
LOD 30, 49, 50, 61, 77, 79, 95, 99, 101, 102
logistics 226–229
long term testing 183
loss on drying 28, 30, 49, 50, 79, 241, 249
lubrication 60, 79, 95, 102, 172
lyophilized 26

manufacturing equipment 1, 12, 74, 117, 152
manufacturing facilities 2, 120, 233
manufacturing process 1, 2, 4, 5, 8, 9, 12, 35, 36, 68–74, 93, 94, 95, 99, 115, 146, 193, 198, 200, 201–204, 206, 207, 214, 220, 221, 230–232, 235, 240, 241, 249
mass balance 24
matrixing 120, 183
mesh size 40, 51
method development 37, 126, 129, 131, 133
method validation 9, 15, 16, 22, 70, 126, 127, 129–131, 133, 246
microbial testing 221, 225
milling 77, 99, 198, 199, 207
mobile phase 17, 19, 32, 37, 126–132
molarity 29, 30, 128

national formulary (NF) 24
negative charge 128
noise level constraint 171
non-polar 127, 128
nonsterile 220

Occupational Safety and Health Administration (OSHA) 26
ointment 32, 198, 199, 202

OOS 18, 60, 82, 91, 107, 196, 205, 224, 225, 243, 244, 246, 249
operational qualification (OQ) 135–139
organic impurities 35, 36
out-of-specification (OOS) results 70, 224, 239, 243
out-of-trend (OOT) results 243

packaging 3–5, 8, 15, 26–28, 58, 70, 74, 75, 81, 88, 91, 95, 97, 98, 106, 115, 155, 184, 187, 193, 195–197
packaging components 218, 220, 224, 236, 238, 240, 248, 249
paddle 19, 23, 39, 41, 43, 45, 48
parameter control tolerance 164
performance qualification (PQ) 69, 72, 135, 136, 138, 139
personal protective equipment (PPE) 194, 237
pH 20–23, 29, 30, 54, 55, 56, 57, 61, 62, 98, 101, 106, 117, 123, 126, 128, 133, 138, 163, 170, 178, 203, 220, 225, 241, 249
pH meter 20–23, 29, 30, 55, 61, 62, 138
pharmaceutical quality 2–4, 232
pharmaceutical quality system 3, 4
Pharmaceutical Regulatory Agency 63
phosphate buffers 54, 129
physical tests 49
pilot batches 69, 189, 191
placebo 37, 38, 130, 131
positive charge 128
primary packaging 197, 199, 207, 210, 211, 213
primary packaging materials (PPM) 197
primary standard 24
process flow diagram 177, 178, 200, 202
process impurities 131
process validation 9, 15, 16, 22, 68–71, 73–75, 87, 91, 93, 95, 97, 115, 135, 190
product grouping 121, 124
project plan 172, 173, 189
prospective validation 70–72, 96, 115
protocols 9, 69, 71, 93, 115, 180, 204, 222, 233, 238

quality assurance 4, 12, 69, 70, 86, 96, 136, 138, 141, 153, 156, 174, 193, 194–209, 212, 214, 215, 216, 217, 218, 224, 228, 231, 232, 234, 237–239, 243, 246
quality control 1, 4–6, 12, 32, 49, 50, 68, 70, 75, 193, 194, 209, 213, 214, 215, 216, 218, 219, 220, 221, 223, 224, 225, 231–241, 243, 246, 247
quality management 2, 66, 215, 218
quantitation limit 126, 130, 132, 133
quantitative determination 25, 29, 30

raw material 2, 5, 8, 13, 49, 51, 53, 55, 57, 59, 61, 73, 74, 76, 77, 97–99, 153, 155, 190, 195, 196, 202, 218, 220, 224, 231, 232, 236, 238, 240, 248
raw material evaluation report 88, 92
reagent grade 21, 28, 57
reagents and catalyst 35
reagents and ligand 35
re-analyze 245
recalibrate 16
re-dilution 244, 245
re-extraction 245
reference standard 24, 25, 27, 29, 31, 190, 221
refractive index 127
refrigeration control 227, 229
regression coefficient 19, 131
re-injection 244, 245
related substances 32, 35, 83, 110, 114, 242
relative humidity 184, 194, 227, 229
relative standard deviation 59, 114, 130
re-sampling 245, 249
residual solvent 35
resolution 130, 153, 159, 245
re-testing 245
retrospective validation 70, 71
revalidation 70, 71, 88, 92, 95, 115
reverse-phase column 127
ribbon blending 87, 91
rinse sampling 123, 125
risk management 2–4, 70, 190, 222, 223, 225
risk management principle 3, 4
robustness 126, 130, 132, 133, 192
role of quality control 5, 218, 219, 221, 223, 225
root cause analysis 3, 246
RPM 19, 20, 39, 41, 45, 60, 77, 99, 202
run time 33, 37, 45, 89, 172

sample preparation 33, 45, 126, 129, 238, 244
sample stock 33
secondary packaging 195, 210, 211
secondary standard 24, 30
self-inspection 4
separation techniques 126
setpoint range 164
shelf life 3, 4, 11, 182, 183, 226
sieving 77, 78, 99
SIM method 187
sinker 41
sitemap 178
soft capsule 55
specificity 123, 126, 130, 133
spectrophotometer 20, 127
stability chamber 185
stability commitment 185
stability indicating method 187
stability storage conditions 184
stability studies 36, 182–188, 244
standard operating procedure 1, 7, 9, 27, 122, 215, 216, 219, 227, 229, 239
standard stock 33, 37, 45
standard working 33, 37, 38, 45
stationary phase 126, 127
sterile testing 220

sterility 11, 14, 199, 207, 222, 234
stress conditions 184
stress testing 183, 187
strip pack 210, 211, 213
sufficient premises 2
sulfonate buffer 129
supply chain 3, 215, 222, 226, 228
swabs and wipes 122–124
syringe filters 129
system suitability 9, 13, 26, 126, 129, 130, 219, 223, 244

tailing factor 130
technology transfer 3, 9, 12, 189–192
temperature mapping 222, 227–229
terminal sterilization 199, 207
tertiary packaging 210, 211
test method 190, 219, 223, 225, 234, 238
test solution 29
testing frequency 184, 187
theoretical plate 130
thin-layer chromatography (TLC) 123
total organic carbon (TOC) 123
toxicology 36

uncoated tablet 44, 53, 54
United States Pharmacopeia (USP) 24
unknown impurity 37
UPLC 126, 130
US Food and Drug Administration 68, 93
USP standard 24
UV detector 127

validated method 32, 187, 246
validation protocol 69, 71, 73, 87, 93, 115, 152, 156, 190, 192
vessel 19, 20, 43, 45, 123
vitro 36, 234
volatile 27
volumetric flask 18, 33, 34, 37, 38, 45, 47, 48
volumetric solution 29, 30

wavelength 45, 126, 127, 129, 132
weighing bottle 49, 61
weight variation 77, 80, 99, 104–106, 202, 241, 249
wet granulation 78, 101
WHO 25, 216

y-intercept 131

Made in the USA
Las Vegas, NV
19 September 2026

141cef81-807f-4ce6-b766-3c3cdf5e08d2R02